Planet3 Wireless, Inc.
PO Box 20063
Atlanta, GA 30325
866-GET-CWNE
www.cwnp.com

The information provided in this manual is provided as-is and is subject to change without notice. All content in this manual are believed to be accurate but are presented without warranty of any kind, express or implied.

LICENSE
PLEASE READ THESE TERMS AND CONDITIONS CAREFULLY BEFORE USING THIS MANUAL ("MATERIALS"). BY USING THE MATERIALS YOU AGREE TO BE BOUND BY THE TERMS AND CONDITIONS OF THIS LICENSE.

NO PART OF THIS DOCUMENT MAY BE REPORDUCED OR TRANSMITTED IN ANY FORM OR BY ANY MEANS, ELECTRONIC OR MECHANICAL, FOR ANY PURPOSE, WITHOUT THE EXPRESS WRITTEN PERMISSION OF PLANET3 WIRELESS.

THE MATERIALS THEREIN ARE PROVIDED "AS IS". PLANET3 WIRELESS MAKES NO REPRESENTATIONS OR WARRANTIES, EITHER EXPRESS OR IMPLIED, OF ANY KIND WITH RESPECT TO THE MATERIALS, THEIR CONTENTS, OR INFORMATION. PLANET3 WIRELESS EXPRESSLY DISCLAIMS ALL WARRANTIES, EXPRESS OR IMPLIED, OF ANY KIND WITH RESPECT TO THE MATERIALS OR THEIR USE, INCLUDING BUT NOT LIMITED TO MERCHANTABILITY AND FITNESS FOR A PARTICULAR PURPOSE. YOU AGREE THAT PLANET3 WIRELESS, ITS DIRECTORS, OFFICERS, EMPLOYEES OR OTHER REPRESENTATIVES SHALL NOT BE LIABLE FOR DAMAGES ARISING FROM THE CONTENT OR USE OF THE MATERIALS. YOU AGREE THAT THIS LIMITATION OF LIABILITY IS COMPREHENSIVE AND APPLIES TO ALL DAMAGES OF ANY KIND, INCLUDING WITHOUT LIMITATION DIRECT, INDIRECT, COMPENSATORY, SPECIAL, INCIDENTAL, PUNITIVE AND CONSEQUENTIAL DAMAGES.

CWNA, CWNI, CWNP, CWSP, CWNE are trademarks of Planet3 Wireless, Inc.

Other trademarks and manufacturer graphics referenced are the property of their respective owners.

Certified Wireless Network Administrator, Revision 2.0, Course Guide
Copyright © 2004 Planet3 Wireless, Inc.
All rights reserved. Printed in the USA.

Acknowledgements

This section acknowledges the efforts of those who contributed significantly to defining Certified Wireless Network Administrator course and making it successful.

Please send specific feedback on the CWNA program or this manual via email to *info@cwnp.com*.

CWNA Core Curriculum Team

Devin Akin - Author
Scott Turner - Graphics, Layout & Formatting, Publishing
Kevin Sandlin - Objectives and Editing

Special thanks go to:

David Westcott - for lab updates from 1.0 to 2.0
Joshua Bardwell - for technical review and various packet analysis content
David Coleman - for beta testing and content review

About the CWNA Course Guide v2

This course guide is intended for use in an authorized CWNA Instructor Led Training (ILT) course. The slide images presented in the course guide match the slides that the CWNT instructor will be presenting during the class. The accompanying text is intended to provide additional information or explanations.

There are 33 chapters followed by a set of lab exercises. The timing of these lab exercises is at the discretion of the CWNT instructor. These lab exercises are intended to give the student hands-on experience with many different brands and types of wireless LAN equipment in a controlled setting. The student should become familiar, comfortable, and proficient at the basic operation of these devices.

At the end of each chapter is a page for writing notes. Students are encouraged to record notes on more complex topics for future reference when preparing for the CWNA certification exam.

Other resources for preparing for the CWNA exam include:

- The CWNA Official Study Guide, ISBN 0-07-222902-0, available at cwne.com
- The CWNA Official Online Practice Test, which includes 450 sample questions delivered in an online practice test engine. A 90-day license is available at cwne.com.

Contents

Chapter 1	Organizations and vendors	1
Chapter 2	Standards and applications	9
Chapter 3	Radio frequency properties and behavior	23
Chapter 4	RF math, FCC power rules, and system operating margin	37
Chapter 5	Spread spectrum technology	55
Chapter 6	Access points, service sets, and the distribution system	63
Chapter 7	Wireless stations, accessory devices, and SOHO networks	73
Chapter 8	Ad Hoc networks	79
Chapter 9	Wireless bridges and workgroup bridges	87
Chapter 10	Antennas, cables, and connectors	97
Chapter 11	Amplifiers, attenuators, splitters, and lightning arrestors	117
Chapter 12	Fresnel zones, free space path loss, interference, fading, and multipath	125
Chapter 13	FHSS	143
Chapter 14	DSSS	151
Chapter 15	OFDM	161
Chapter 16	PBCC	167
Chapter 17	Channels, data rates, ranges, and comparisons	175
Chapter 18	Channel reuse and mixed mode	189
Chapter 19	Arbitration, RTS/CTS, carrier sense and hidden nodes	197
Chapter 20	Beacons, probes, and scanning	211
Chapter 21	Authentication, association, and roaming	217
Chapter 22	PoE and 802.3af	231
Chapter 23	Fragmentation and acknowledgments	239
Chapter 24	Standards-based roaming	251
Chapter 25	Bandwidth control, network management, and AAA	257
Chapter 26	Routers, gateways, switches, and VoWiFi	265
Chapter 27	Power management	279
Chapter 28	Security mechanisms and common threats	287
Chapter 29	802.1x/EAP, TKIP, WPA, 802.11i, and AAA	299
Chapter 30	Baseline practices, common security solutions, and regulations	313
Chapter 31	Site surveys: administrative perspective	331
Chapter 32	Site surveys: technical perspective	353
Chapter 33	WEP	385
Lab 1	Infrastructure mode connectivity	399

Lab 2	Infrastructure mode throughput analysis	**407**
Lab 3	Ad Hoc connectivity and throughput analysis	**415**
Lab 4	Cell-sizing and ARS in infrastructure mode	**421**
Lab 5	Co-channel and adjacent channel interference	**431**
Lab 6	Basic security features	**439**
Lab 7	Dynamic WEP keys using 802.1x/EAP and RADIUS	**447**
Lab 8	Wireless VPN using PPTP tunnels and RADIUS	**459**
Appendix A	Site survey questionnaire	**465**
Appendix B	Access point form	**473**
Appendix C	Bridge form	**481**
Appendix D	Workgroup bridge form	**489**
Appendix E	AirMagnet – Performing the site survey with AirMagnet	
Appendix F	AirMagnet – Managing 802.11g with AirMagnet Trio	
Appendix G	Wildpackets – Converting signal strength percentage to dBm values	

Chapter 1
Organizations and Vendors

Objectives

Upon completion of this chapter you will be able to:

- Understand the roles of the organizations and regulations that have a significant impact on the wireless LAN industry
- Discuss the leading market vendors in the SOHO and Enterprise market spaces

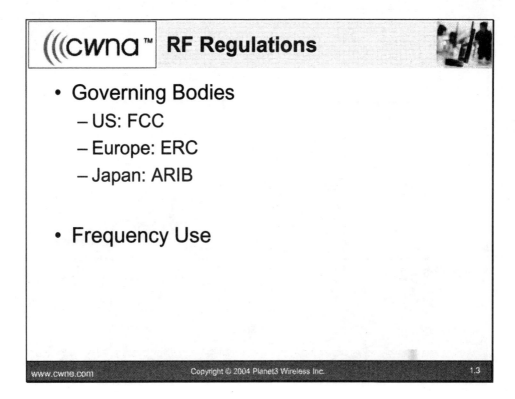

The allocation of radio frequencies is usually overseen by a governing body:

- U.S.: FCC (Federal Communications Commission)
- Europe: ERC (European Radiocommunications Committee)
- Japan: ARIB (Association of Radio Industries and Businesses)

These bodies determine who may use which frequencies. Significant effort is placed into getting the various organizations and governing bodies to agree. The 2.4 GHz range is similarly specified nearly worldwide.

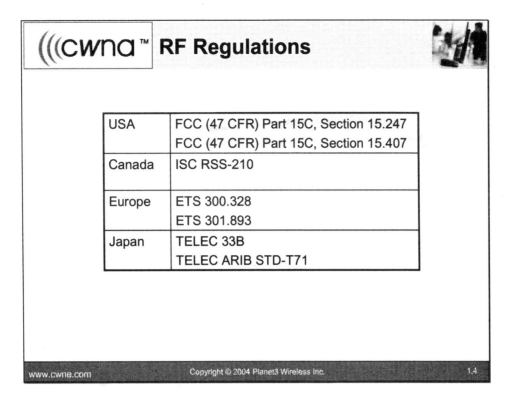

Because there are hundreds of agencies worldwide with regulatory control over RF frequencies, allocation of the RF spectrum may differ from country to country.

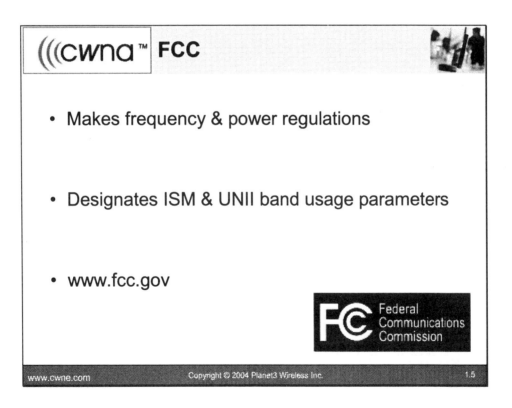

The FCC makes the regulations regarding RF frequency use and power output limits for the United States. The FCC designates ISM and UNII band usage parameters such as frequency bands used, power output limits, license-free use of the frequency bands, and the type of RF transmissions (e.g., spread spectrum technology). Complete information about the FCC, including regulations, can be found at the FCC web site - www.fcc.gov.

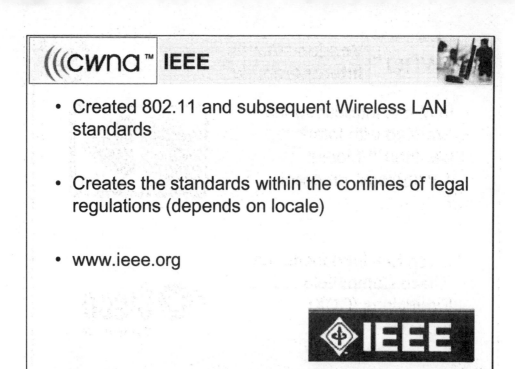

The IEEE (Institute of Electrical and Electronics Engineers) creates the industry standards regarding wireless LANs in the United States and in many parts of the world. The IEEE created the 802.11 standard and subsequent wireless LAN standards (a, b, g, etc.). The IEEE creates the standards within the confines of legal regulations, depending on the local governing body. The IEEE web site is www.ieee.org.

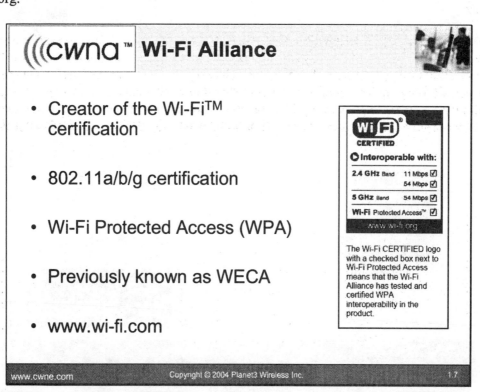

The Wi-Fi Alliance, formerly known as WECA, or the Wireless Ethernet Compatibility Alliance, is the creator of the Wi-Fi™ certification for wireless LAN hardware interoperability. This certification applies to all 802.11a/b/g compatible equipment, which must undergo third party testing to ensure its compatibility. The Wi-Fi Alliance also created Wi-Fi Protected Access (WPA) Certification for basic wireless LAN security. The Wi-Fi Alliance's web site is at www.wi-fi.com.

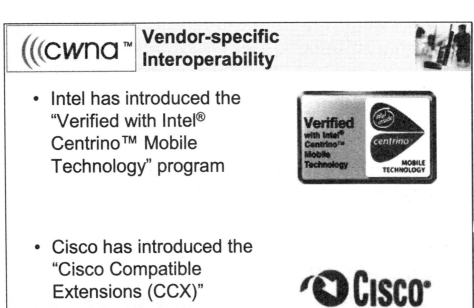

Intel has introduced the "Verified with Intel® Centrino™ Mobile Technology" program. Intel's goal is to eventually put the 802.11-based radio into the main CPU. Currently, the Centrino chipset is an additional chipset on the motherboard. Intel's reach into the PC market will allow Intel to become a major force in the wireless LAN market, assuming their equipment is interoperable with devices from other manufacturers.

Cisco has introduced the "Cisco Compatible Extensions (CCX)" program. Cisco seeks to control the wireless infrastructure, and one of the ways of accomplishing this goal is to release their newly developed technology to partners for the purpose of adding the technology to their wireless LAN clients and RADIUS servers. By maintaining control of the infrastructure device code (access points, bridges, etc.), Cisco will allow all partners implementing CCX to work with Cisco infrastructure devices – and no others – using these advanced features. The CCX program offers advantages and disadvantages to the consumer.

Leading Wi-Fi Manufacturers

- SOHO
    - Buffalo Technologies
    - DLink
    - SMC
    - Linksys
    - Netgear
    - Microsoft
    - Belkin

Leading Wi-Fi Manufacturers

- Enterprise
    - Cisco
    - Symbol
    - Proxim
    - Intermec
    - Colubris
    - Enterasys
    - HP

Chapter 2
Standards and Applications

Objectives

Upon completion of this chapter you will be able to:

- Describe IEEE 802.11 series of standards that are relevant in today's market
- Describe appropriate applications of wireless LAN technology in access and distribution roles

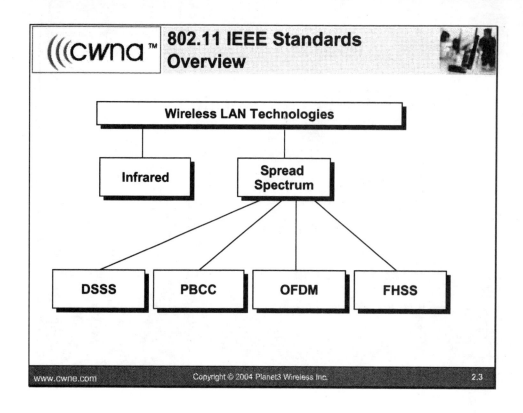

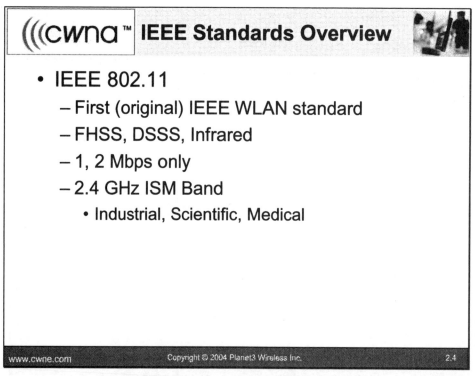

The IEEE 802.11 is the first (original) IEEE wireless LAN standard. The 802.11 standard was ratified 1997, and amended in 1999. The 802.11 standard specifies functionality for frequency hopping (FHSS), direct sequence (DSSS), and Infrared technologies. The 802.11 standard only specifies data transfer speeds of 1 and 2 Mbps, and applies only to the 2.4 GHz ISM (Industrial, Scientific, Medical) band.

The 802.11 standard specifies Point to MultiPoint (PtMP) output power maximums of:

- 1 Watt @ Intentional Radiator (IR)
- 4 Watts @ Equivalent Isotropically Radiated Power (EIRP)

 **IEEE Standards Overview**

- IEEE 802.11b
  - DSSS Only
  - 1, 2, 5.5, 11 Mbps rate support mandatory
  - Backward compatible with 802.11 for DSSS only
  - 2.4 GHz ISM Band

The IEEE 802.11b standard was ratified in 1999, and amended in 2001. 802.11b specifies DSSS technology only, and mandates support for data rates of 1, 2, 5.5, and 11 Mbps. 802.11b products are backward compatible with 802.11 for DSSS only, and work only in the 2.4 GHz ISM Band.

802.11b also specifies Point-to-Multipoint (PtMP) output power maximums:

- 1 Watt @ Intentional Radiator (IR)
- 4 Watts @ Equivalent Isotropically Radiated Power (EIRP)

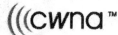

IEEE Standards Overview

- IEEE 802.11a
  - OFDM Only
  - 6, 9, 12, 18, 24, 36, 48, 54 Mbps
  - 6, 12, 24 Mbps support mandatory
  - 5 GHz UNII bands (UNII-1, UNII-2, UNII-3)
    - Unlicensed National Information Infrastructure

The IEEE 802.11a standard was ratified in 1999 for use of OFDM technology only. 802.11a allows for support of date rates of 6, 9, 12, 18, 24, 36, 48, 54 Mbps, and mandates support for data speeds of 6, 12, 24 Mbps.

802.11a utilizes the 5 GHz UNII bands (UNII-1, UNII-2, and UNII-3). "UNII" stands for Unlicensed National Information Infrastructure, and is divided into three separate frequency bands:

- UNII-1 (5.15-5.25 GHz), 40mW PtMP@IR
- UNII-2 (5.25-5.35 GHz), 200mW PtMP@IR
- UNII-3 (5.725-5.825 GHz), 800mW PtMP@IR

IEEE Standards Overview

- **IEEE 802.11g**
  - OFDM, DSSS support mandatory
  - PBCC support optional
  - 6, 12, 24 Mbps OFDM mandatory
  - 6, 9, 12, 18, 24, 36, 48, 54 Mbps OFDM supported
  - 2.4 GHz ISM band

The IEEE 802.11g standard was ratified in 2003, and mandates support for both OFDM, DSSS technologies, with optional support for PBCC (Packet Binary Convolutional Coding).

802.11g requires the following:

- 5.5, 11 Mbps DSSS mandatory for backward compatibility with 802.11b
- 6, 12, 24 Mbps OFDM mandatory
- 6, 9, 12, 18, 24, 36, 48, 54 Mbps OFDM supported
- Utilization of the 2.4 GHz ISM band
- Point-to-Multipoint output power maximums:
  - 1 Watt @ Intentional Radiator (IR)
  - 4 Watts @ Equivalent Isotropically Radiated Power (EIRP)

IEEE Standards Overview

- IEEE 802.11e (QoS) - draft
  - EDCF (Enhanced Distributed Coordination Function)
  - HCF (Hybrid Coordination Function)

IEEE Standards Overview

- IEEE 802.11f (IAPP)
  - A "Recommended Practice" for interoperability between vendors for AP-to-AP roaming
  - Covers the Inter Access Point Protocol (IAPP)

IEEE 802.11f (IAPP) is not a standard, but rather a "recommended practice" for interoperability between vendors for access point to access point roaming. 802.11f covers the Inter Access Point Protocol (IAPP), communication over the Distribution System (DS), exchange of information between access points about mobile clients, maintenance of bridge forwarding tables, and securing communications between access points.

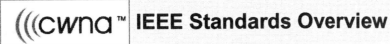

IEEE Standards Overview

- IEEE 802.11h
  - Spectrum-managed 802.11a technology
  - Transmit Power Control (TPC)
  - Dynamic Frequency Selection (DFS)
  - Mandatory in Europe

The IEEE 802.11h standard specifies spectrum-managed 802.11a technology, and includes the following specifications:

- Support for transmit power control (TPC), in which the power output is managed by the access point and clients to avoid interfering with other nearby systems unnecessarily.
- Support for dynamic frequency selection (DFS), in which the access point chooses the channel to avoid interference with other systems

802.11h is mandatory in Europe, according to the European Telecommunications Standards Institute (ETSI).

IEEE Standards Overview

- IEEE 802.11i (Security)
  - 802.1x/EAP, AES, and others
  - WPA v1.0 is stop-gap standard from the Wi-Fi Alliance
  - WPA v2.0 will be fully 802.11i compliant

The IEEE 802.11i standard, which is still in development, is specific to wireless LAN security. 802.11i specifies 802.1x/EAP, AES, and other technologies introduced to enhance security beyond 802.11.

WPA v1.0 is stop-gap standard from the Wi-Fi Alliance and an interim snapshot of the 802.11i standard. WPA v2.0 will be fully 802.11i compliant according to the Wi-Fi Alliance.

Multi-Standard chip features include:

- 802.11a/b/g
- 802.11i (802.1x/EAP, WPA, AES)
- 802.11e (EDCF/HCF)
- 802.11f (Roaming)
- 802.11h (TPC/DFS)
- Enhanced Low Power (ELP)

Multi-Standard chips are available from:

- Texas Instruments
- Globespan Virata (products formerly owned by Intersil)
- Atheros
- Broadcom
- Intel
- Lucent
- Cisco

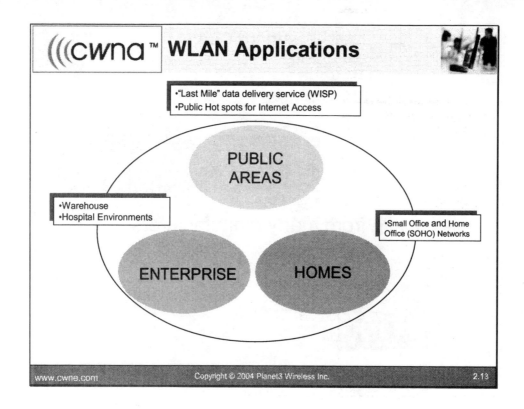
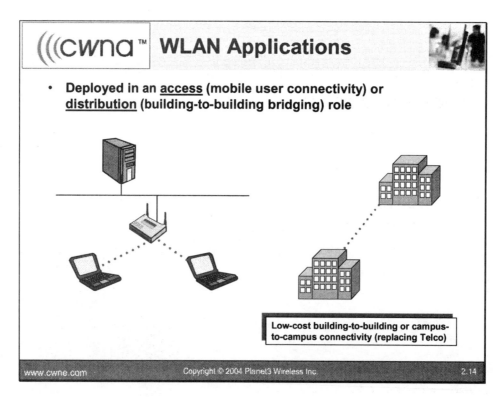

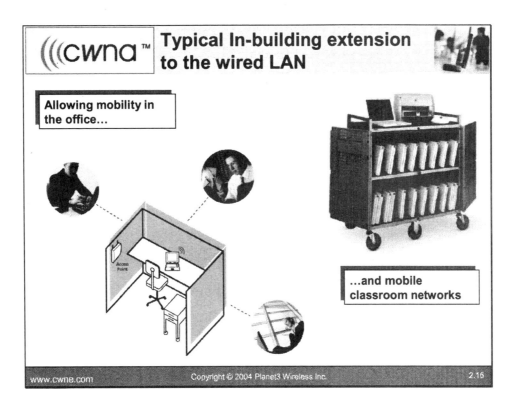

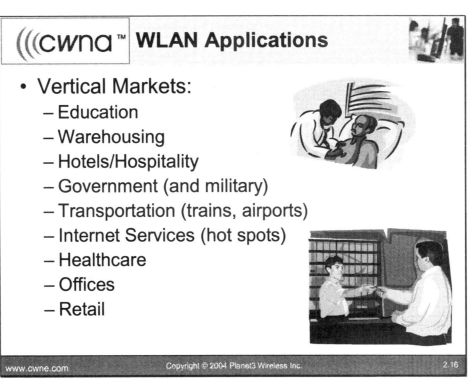

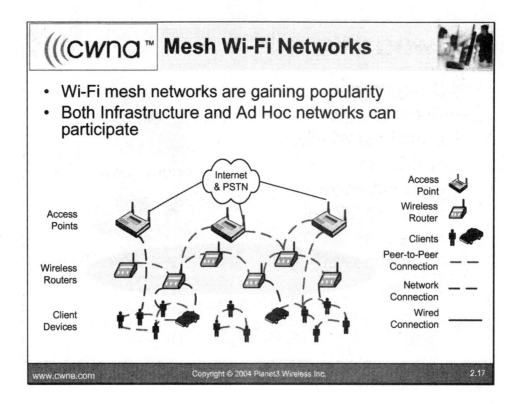

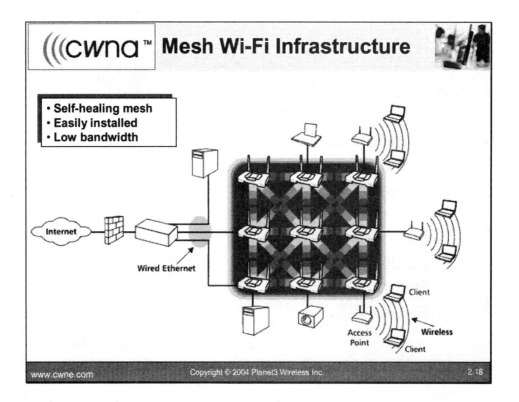

Hot Spot Providers

- Boingo.com
- CometaNetworks.com
- Deep Blue Wireless
- GRIC.com
- HotSpotList.com
- HOTSPOTZZ.com
- iPass.com
- StayOnline.net
- Verizon
- Hotspot.Toshiba.com
- SurfandSip.com
- Wayport.com
- Starbucks.com
- WiFinder.com
- T-Mobile.com/hotspot/
- McDWireless.com
- Fatport.com
- Truckstop.net

Chapter 3
Radio Frequency Properties and Behavior

 **Objectives**

Upon completion of this chapter you will be able to:

- Define Radio Frequency
- Identify and define the basic properties of Radio Frequency (RF)
- Explain basic RF behavior

Radio Frequency

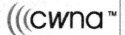

- **RF Properties**
  - Wavelength
  - Frequency
  - Amplitude
  - Phase
  - Amplification (Gain)
  - Attenuation (Loss)

- **RF Behavior**
  - Reflection
  - Refraction
  - Diffraction
  - Scattering
  - Absorption
  - VSWR
  - Return Loss

RF Basics

- **Definition of RF in a Wireless LAN:** High Frequency (300 kHz – 300 GHz) electromagnetic waveforms that are passed along a copper conductor and then radiated into the air via an antenna element

- All Electromagnetic (or EM) signals have certain basic characteristics
  - Wavelength
  - Frequency

Wavelength

- **Wavelength** is the length (distance) of one oscillation, or cycle, of the signal
  - Wavelength of RF signals are usually measured in **centimeters**
  - The symbol λ (lambda) stands for Wavelength

Longer Wavelength

Shorter Wavelength

Frequency

- **Frequency** is the number of times per second that the signal oscillates
  - Frequency is measured in **Hertz (Hz)**
  - One cycle per second equals one Hertz
  - The letter *f* stands for frequency

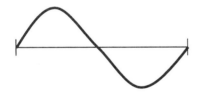
Lower Frequency

Higher Frequency

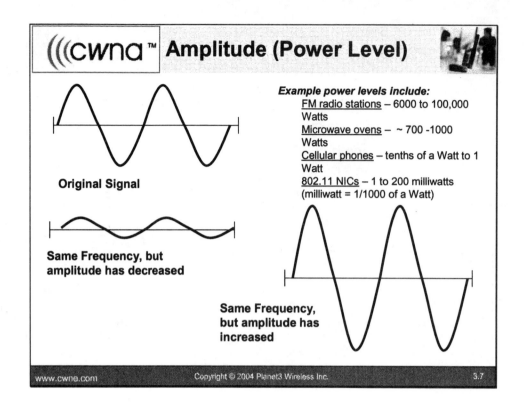

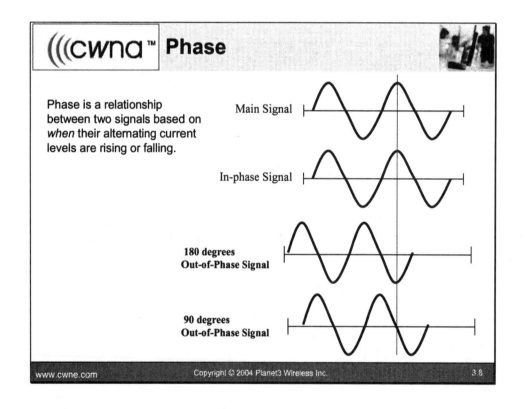

Wavelength & Frequency

- Wavelength and Frequency are <u>interchangeable</u> and related through these formulas:

$$f = \frac{c}{\lambda} \qquad \lambda = \frac{c}{f}$$

- f is frequency; λ is wavelength; c is the speed of light—about 300,000 km/s

This formula shows that, as frequency goes up, wavelength goes down, and vice versa. This concept is true only for energy with a fixed speed, such as electromagnetic energy.

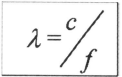

Wavelength Calculation

- 2.45 GHz = 4.82 inches (12.24 cm)

- 5.775 GHz = 2.04 inches (5.19 cm)

- Formulas to calculate RF wavelengths:
  - Wavelength (inches) = 11.811/Frequency (GHz)
  - Wavelength (cm) = 30/Frequency (GHz)

802.11a (5 GHz) wavelengths do not travel as far as 802.11b (2.4 GHz) during a single cycle. Higher frequencies have shorter ranges at equal power.

Metric Prefix Refresher

- Metric prefixes are often used when describing electromagnetic signals
- Here are some common metric prefixes

Prefix	Example	Meaning	Multiplier
µ-	µm	micro	1/1,000,000
n-	nm	nano	1/1,000,000,000*
k-	km, kHz	kilo	1,000
M-	MHz	mega	1,000,000
G-	GHz	giga	1,000,000,000

* So one nm would be one one-billionth of a meter.

Some examples of wavelengths and frequencies include visible light, which has wavelengths from 400 nm to 700 nm, and cordless phones, which typically use frequencies around 900 MHz or 2.4 GHz. FM radio stations use frequencies from 88 to 108 MHz. The 802.11 series of standards specify use of frequencies around 2.4-2.5 GHz and 5-6 GHz.

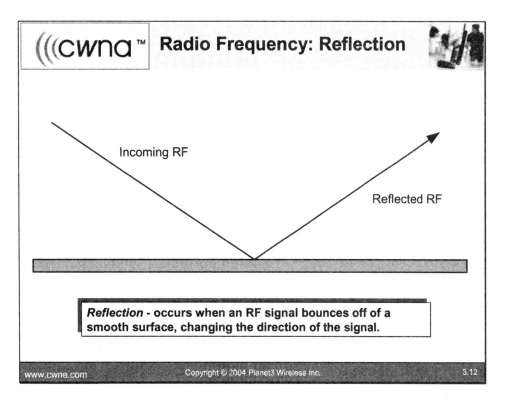

Reflection occurs when a propagating electromagnetic wave impinges upon an object that has very large dimensions relative to the wavelength of the propagating wave.

Reflections occur from the surface of the earth, buildings, walls, and many other obstacles. If the surface is smooth, the reflected signal may remain intact, though there is some loss due to absorption and scattering of the signal.

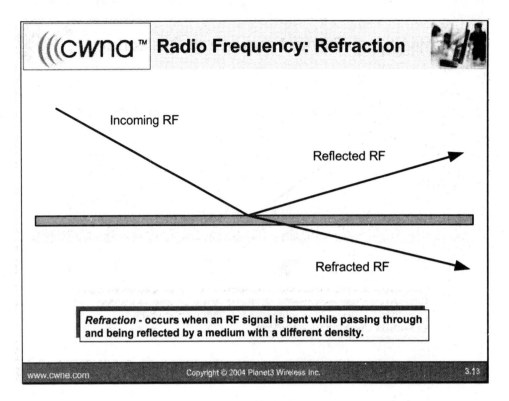

Refraction describes the bending of a radio wave as it passes through a medium of different density. As an RF wave passes into a denser medium (like a pool of cold air hovering over a valley) the wave will be bent such that its direction changes. When passing through such a medium, some of the wave will be reflected away from the intended signal path, and some will be bent through the medium in another direction. Refraction can become a problem for long distance RF links. As atmospheric conditions change, the RF waves may change direction, diverting the signal away from the intended target.

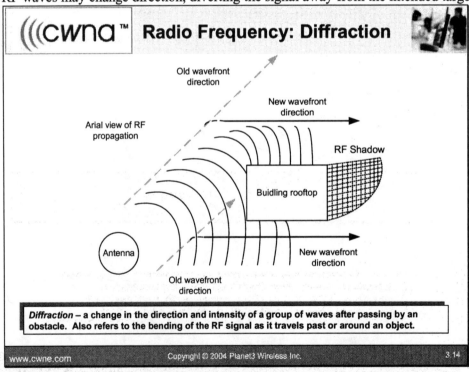

Diffraction occurs when the radio path between the transmitter and receiver is obstructed by a surface that has sharp irregularities or an otherwise rough surface. At high frequencies, diffraction, like reflection, depends on the geometry of the obstructing object and the amplitude, phase, and polarization of the incident wave at the point of diffraction.

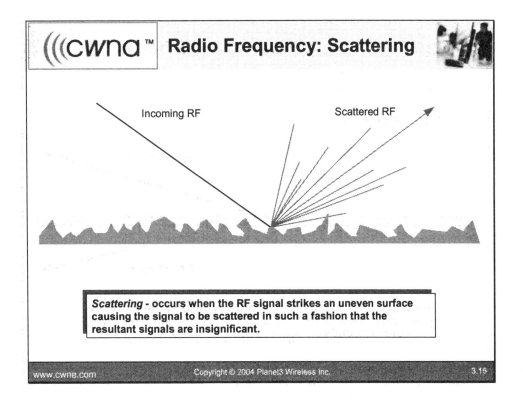

Scattering occurs when the medium through which the wave travels consists of objects with dimensions that are small relative to the wavelength of the signal, and the number of obstacles per unit volume is large. Scattered waves are produced by rough surfaces, small objects, or by other irregularities in the signal path. Some outdoor examples of objects that can cause scattering in a mobile communications system include foliage, street signs, and lampposts.

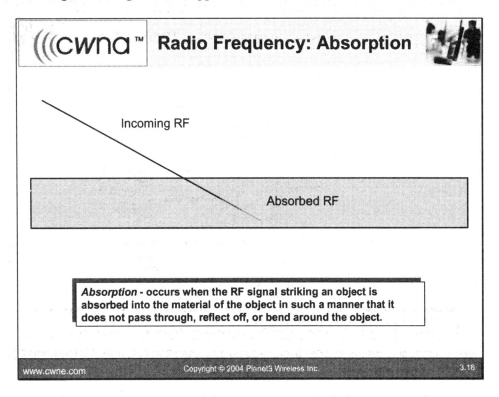

Absorption occurs when the RF signal strikes an object and is absorbed into the material of the object in such a manner that the RF signal does not pass through, reflect off, or bend around the object.

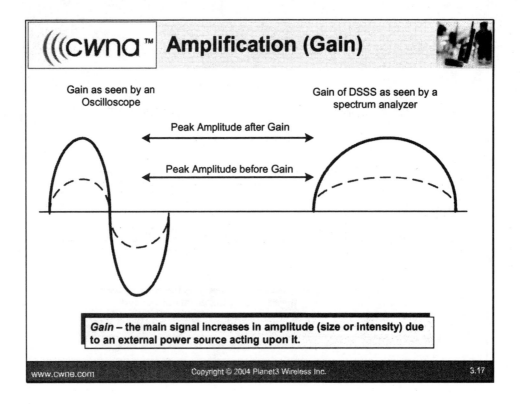

Gain is usually an active process in which an external power source, such as an RF amplifier, is used to amplify the signal. Passive processes can also cause gain. Reflected RF signals can combine with the main signal to increase the main signal's strength. Antennas can be used to focus a signal in one particular direction. Typically, more power is better, but there are cases, such as when a transmitter is radiating power very close to the legal power output limit or when there are other transmitters nearby, where added power would be a serious problem due to legal issues or interference problems.

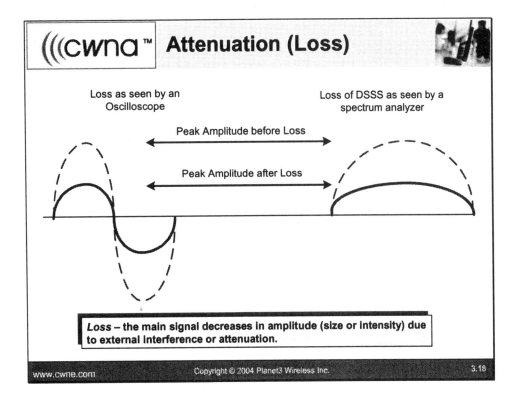

Many things can cause RF signal loss, both while the signal is still in the cable as a high frequency AC electrical signal and when the signal is propagated as radio waves through the air by the antenna.

Resistance of cables and connectors cause loss due to the converting of the AC signal to heat. Impedance mismatches in the cables and connectors can cause power to be reflected back toward the source, which can cause signal degradation. Objects directly in the propagated wave's transmission path can absorb, reflect, or destroy RF signals.

Loss can be intentionally injected into a circuit with an RF attenuator. RF attenuators are accurate resistors that convert high frequency AC to heat in order to reduce signal amplitude at that point in the circuit.

Being able to measure and compensate for loss in an RF connection or circuit is important because radios have a receive sensitivity threshold. A sensitivity threshold is defined as the point at which a radio can clearly distinguish a signal from background noise. Since a receiver's sensitivity is finite, the transmitting station must transmit a signal with enough amplitude to be recognizable at the receiver. If losses occur between the transmitter and receiver, the problem must be corrected either by removing the objects causing loss or by increasing the transmission power.

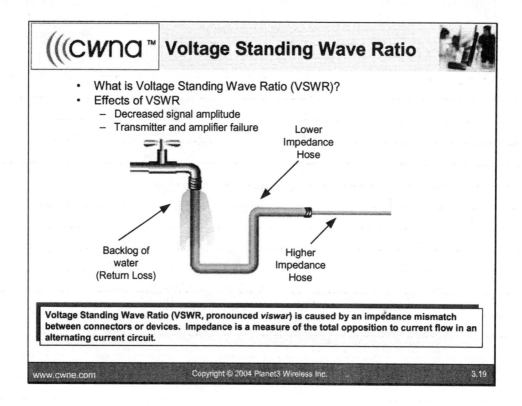

Antennas are usually located some distance from the transmitter and require a series of cables and connectors to transfer power from the transmitter to the antenna. If this series of cables has *no* loss, and matches *both* the transmitter output impedance *and* the antenna input impedance, only then will maximum power be delivered to the antenna. In this case, the VSWR will be 1:1 and the voltage and current will be constant over the whole length of the series. Any deviation from this situation will cause a "standing wave" of reflected voltage to exist on the line. Most manufacturers of wireless LAN hardware provide specifications that say their equipment is specified at a VSWR of 1.5:1 or better. Military requirements are typically 1.1:1 or better.

It is *imperative* that all devices, connectors, and cables in the transmission path have the same impedance (as closely as possible). Most devices used with wireless LANs have 50-ohm impedances. An ohm is defined as a unit of measure for resistance to current flow in an electrical circuit.

WARNING: Do not power up a transmitter without a termination (load) or an antenna attached – sustained VSWR at high power could damage the transmitter.

There is always a given amount of 'return loss' (caused by VSWR) due to imperfect impedance matching between components. Return loss is the term used to describe the forward power loss due to reflected power caused by impedance mismatches. The goal when installing cables and connectors is to match them as closely as possible so as not to introduce any unnecessary return loss.

Chapter 4
RF Math, FCC Power Rules, and System Operating Margin

Objectives

Upon completion of this chapter you will be able to:

- Define decibels
- Explain why decibels are used
- Explain Receiver Sensitivity
- Explain "Intentional Radiator" and "EIRP"
- Explain FCC regulations regarding output power maximums in ISM and UNII bands
- Perform RF Link Budget Calculations

decibels (dB)

- decibels are a unit of measure for a <u>CHANGE</u> in power

- To calculate the decibel difference between two signals, use the following formula:

$$dB = \log\left(\frac{received\ power}{transmitted\ power}\right) * 10$$

A decibel is different from typical units of measurement in that it measures the relative power (the difference or ratio) of two different signals. Typical measurements measure the absolute power of a single quantity.

For the normalization of the decibel, remember that Watts is the unit of measure for absolute power and milliWatts (mW) is a common unit of measure when working with radio frequency.

Decibels are very useful for measuring RF signal strength. Sometimes, we want to measure an absolute power level. In order to use decibels to measure an absolute amount of power, it is common to normalize (reference against) the decibel measurements to a known value.

We can arbitrarily decide that the reference signal is 1 mW.

We can measure the dB difference between a signal and 1 mW.

We can now say "X dB referenced against 1 mW" = dBm.

dBm is now our logarithmic unit of measure for absolute power.

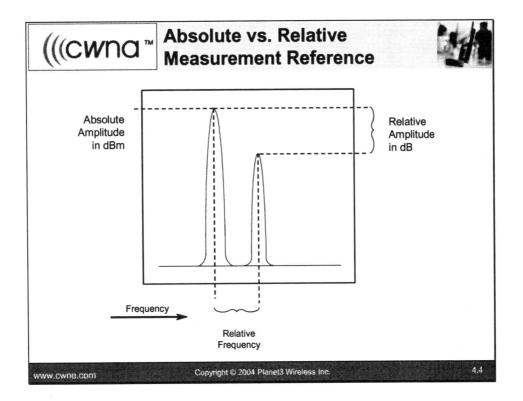

ABSOLUTE units of power measurement:

- Watts (W)
- Milliwatts (mW)
- dBm

RELATIVE units of power measurement (change in power):

- dB
- dBi
- dBd

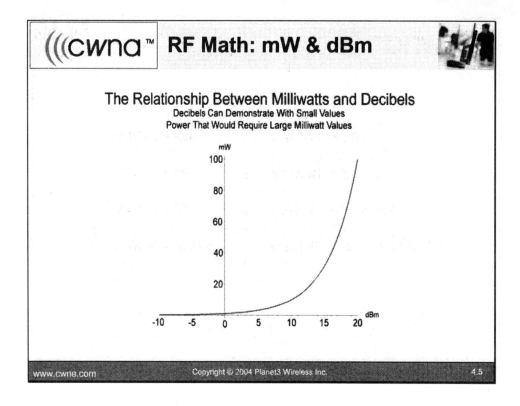

Rather than measuring the absolute amount of power lost, RF engineers measure the relative strength of the transmitted and received signals. For example, one might say, "the received signal is 1/1,000 as strong as the transmitted signal," or "the received signal is 1/1,000,000 as strong as the transmitted signal."

In order to simplify this measurement, RF engineers use a special unit of measurement that is specifically designed to measure the relative strengths of two signals. This unit of measurement is known as the bel (after Alexander Graham Bell).

A more common unit is the decibel (dB), which is equal to 1/10 of a bel. The decibel as a unit of measurement was originally created by the phone company to measure voltage loss on phone lines. The decibel is a logarithmic unit of measure.

RF Math: Relationship of 10s & 3s

1 mW = 0 dBm	1 mW = 0 dBm
10 mW = 10 dBm	2 mW = 3 dBm
100 mW = 20 dBm	4 mW = 6 dBm
1W = 1000 mW = 30 dBm	8 mW = 9 dBm

Reminders for the relationship of the 10s and the 3s

- -3 dB = half the power in mW
- +3 dB = double the power in mW
- -10 dB = one tenth the power in mW
- +10 dB = ten times the power in mW

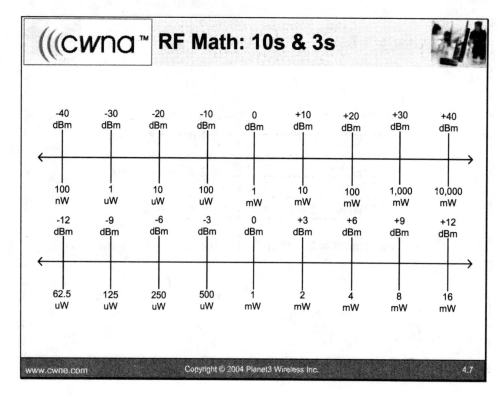

Decibels are logarithmic, meaning that, if the signal gets 10 times weaker, the decibel rating decreases by 10.

- Signal is 0 times weaker = 0 dB
- Signal is 10 times weaker = -10 dB
- Signal is 100 times weaker = -20 dB
- Signal is 1,000 times weaker = -30 dB
- Signal is 10,000 times weaker = -40 dB
- Signal is 100,000 times weaker = -50 dB
- Signal is 1,000,000 times weaker = -60 dB
- Signal is 10,000,000 times weaker = -70 dB
- Signal is 100,000,000 times weaker = -80 dB

Just a few decibels can make a big difference in signal strength. A doubling or halving of signal strength corresponds to about 3 dB difference.

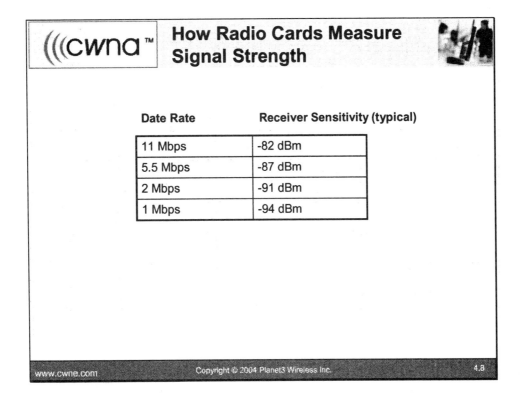

Wireless cards typically indicate signal strength using dBm units instead of mW. dBm measures the actual signal strength relative to 1 mW.

- 0 dBm = 1 mW (reference value)
- -10 dBm = 0.1 mW
- -20 dBm = 0.01 mW
- -30 dBm = 0.001 mW

dBm are also used to measure receiver sensitivity, which calculates the weakest signal the card can receive. This signal strength is usually different for different data rates.

Receive Signal Strength Indicator (RSSI) Value

- The RSSI value is a relative value of the <u>strength</u> and <u>quality</u> of an RF signal being received by the antenna

- Manufacturers can implement this any way they choose
    - RSSI will not mean the same thing across multiple vendors

> Joe Bardwell has written a whitepaper called, "Converting Signal Strength Percentage to dBm Values" that thoroughly explains RSSI values.

The RSSI value is a relative value of the strength and quality of an RF signal being received by the antenna. The RSSI value is referenced in the 802.11 standard, but not defined. Manufacturers can implement this any way they choose. RSSI will not mean the same thing across multiple vendors.

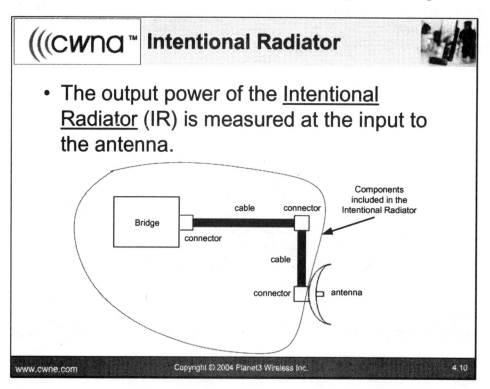

The FCC defines the Intentional Radiator (IR) as a device that transmits RF energy for the purpose of communication. The output power of the IR is measured at the input to the antenna, meaning that the IR measurement is a consolidation of the transmitter, connectors, cables, amplifiers, attenuators, splitters, and all other devices, but NOT the antenna gain.

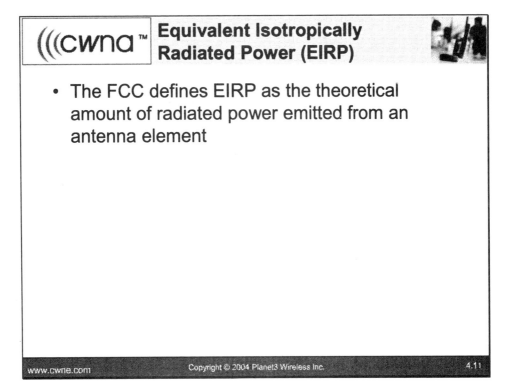

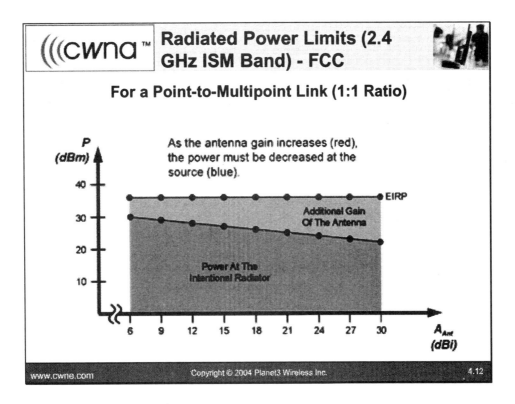

Reference Point: 1 Watt (30dBm) at the IR with 6 dBi antenna gain produces 4 Watts (36 dBm) EIRP Point-to-Multipoint (PtMP). Starting at the maximum as a reference point, increases in antenna gain must result in equal (1:1) decreases in IR power.

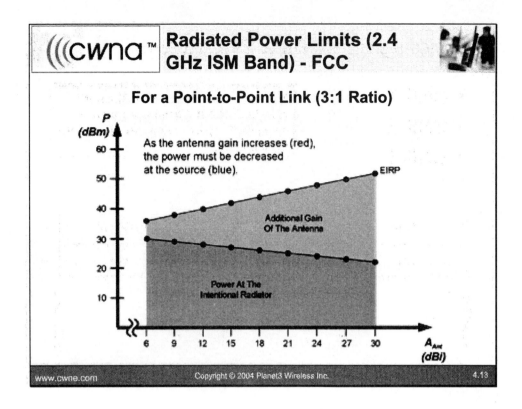

Point-to-Point (PtP): Starting at the maximum as a reference point, an increase of 3 dB in antenna gain must result in a decrease of 1 dB at the IR.

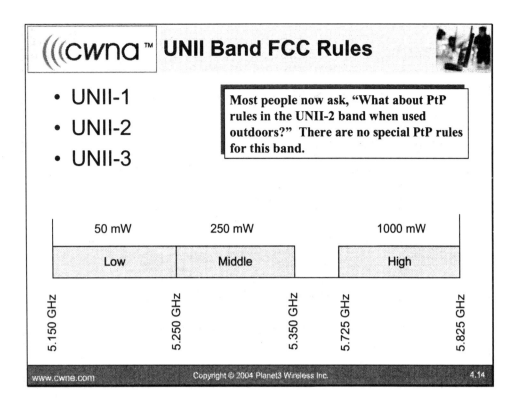

UNII-1 Rules

- Fixed Omni antennas only
- 50mW PtMP@IR
- 6 dBi antenna gain allowed
- Indoor use only

UNII-2 Rules

- Detachable antennas allowed
- 250mW PtMP@IR
- 6 dBi antenna gain allowed
- Indoor/Outdoor use

UNII-3 Rules

- Detachable antennas allowed
- 1 Watt PtMP@IR
- 6 dBi antenna gain allowed
- Outdoor use only
- Special PtP Power output rule

The UNII-3 band (5.725-5.825 GHz) allows an IR power of 1 Watt, but allows antenna gain to boost EIRP to 200 Watts (a gain of 23 dBi) without backing off on the transmission power when using directional antennas in a PtP configuration. After 23 dBi of antenna gain, there is a 1:1 ratio of antenna gain to intentional radiator back off.

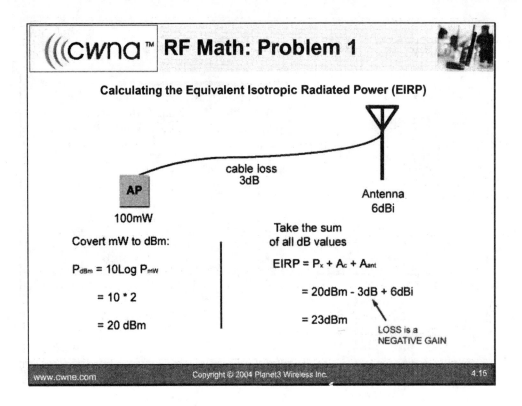

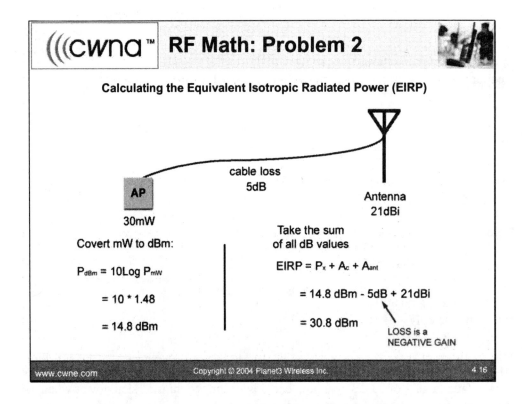

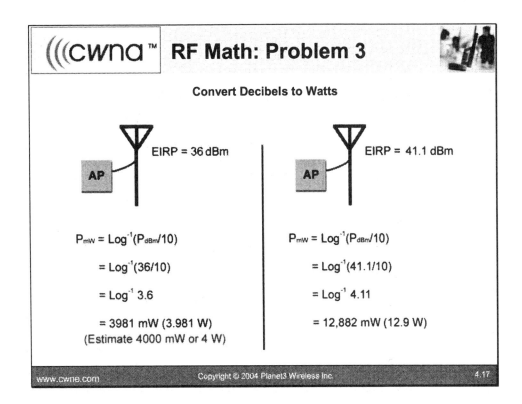

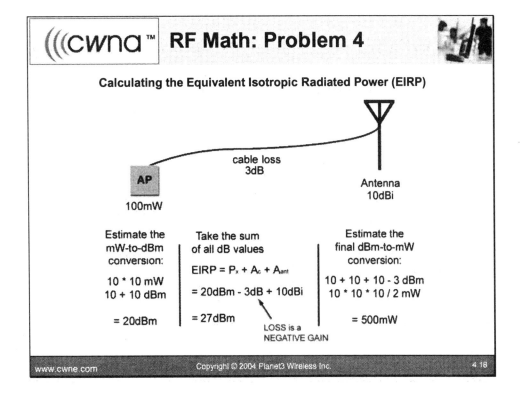

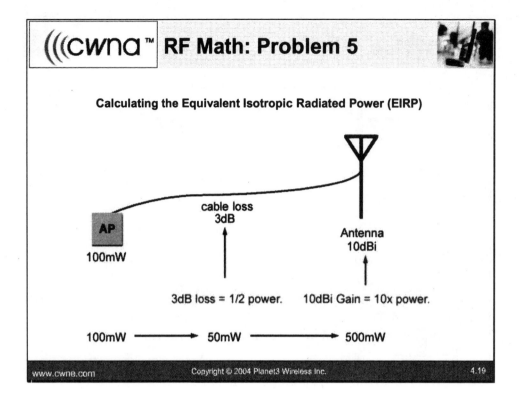

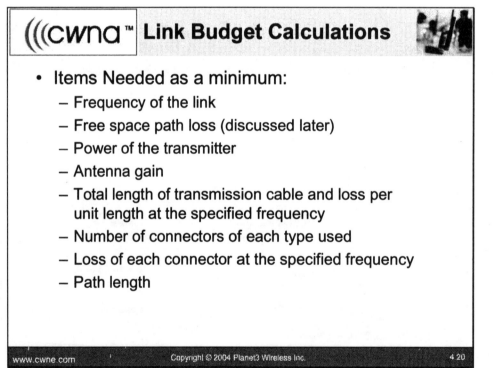

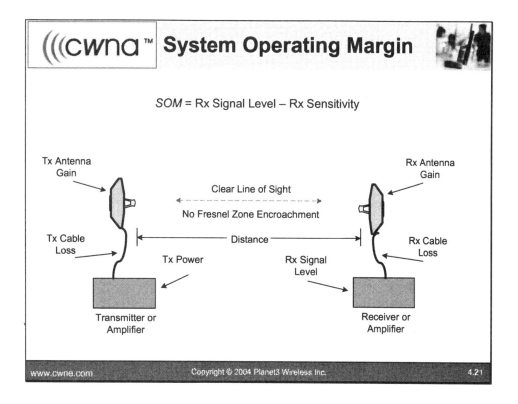

System Operating Margin (SOM) is also known as Fade Margin. SOM is the difference between the signal that a radio is actually receiving versus what it needs for good data recovery (i.e. receiver sensitivity). In order to "receive" an RF signal, the receiver must be able to distinguish the signal from background "noise" or RF interference. Receivers come in varying levels of sensitivity. Better (more sensitive) receivers are able to distinguish weaker and more degraded signals. Some receivers have two or more antennas in order to increase their sensitivity. A higher SOM is always better for higher availability of a link. 10dB is the industry standard minimum, but SOMs of 20dB or higher are considered excellent.

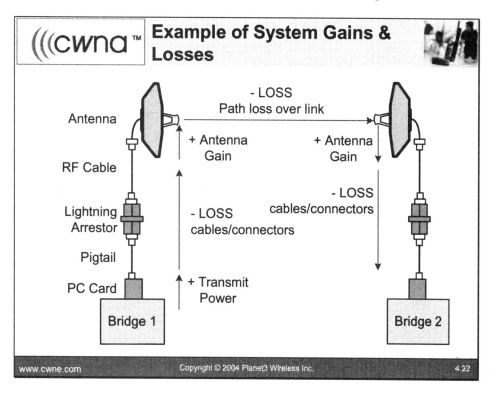

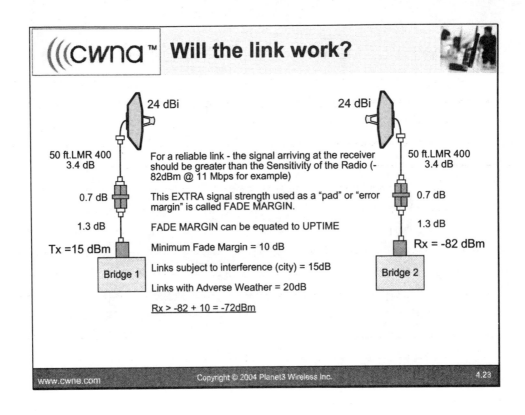
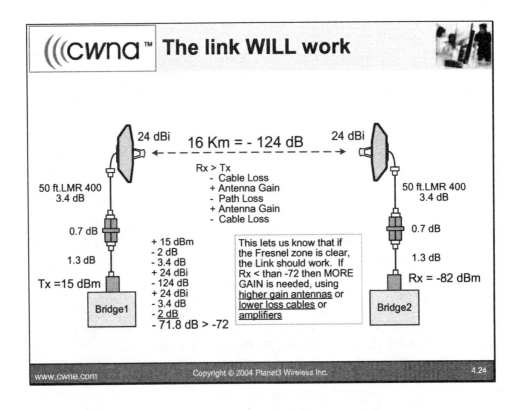

Chapter 5
Spread Spectrum Technology

Objectives

Upon completion of this chapter you will be able to:

- Explain the difference between Narrowband and Spread Spectrum RF communication
- Define the frequency bands allocated by the FCC for spread spectrum technology use in the United States
- Explain how Spread Spectrum communication works
- Define Frequency Ranges and Channels used for Wireless LANs and their importance
- Explain the importance of chipping code, processing gain, and spreading

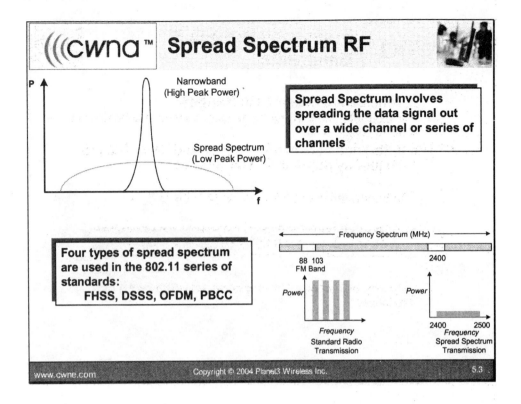

DSSS technology was originally developed by the military as a way of hiding data near the noise floor, which made it more difficult for the enemy to intercept communications.

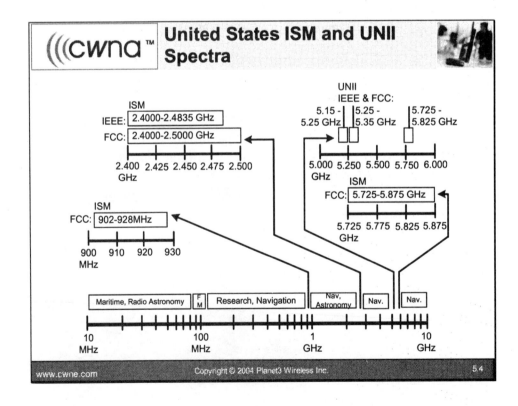

 ## Frequency Ranges & Bandwidth

- Frequencies are assigned in **Ranges**
  - For example, the 2.4GHz range extends from 2.4000 GHz to 2.4835 GHz
- The difference between the upper and lower bounds of a frequency range is its **Bandwidth**

  - The bandwidth of the 2.4 GHz range in the U.S. is:

  $$2.4835 - 2.4000 = 0.0835 \text{ GHz} = 83.5 \text{ MHz}$$

  - Generally, higher bandwidth ranges can support more data throughput

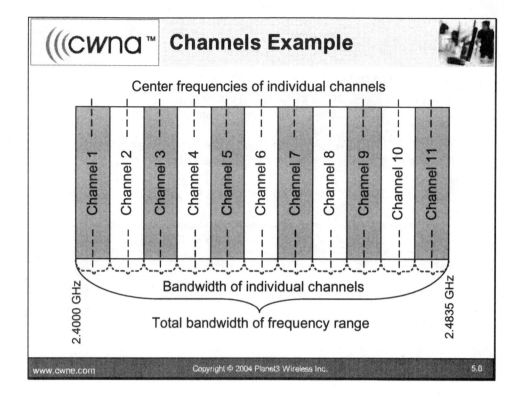

It is common to sub-divide a frequency range into sub-ranges, known as channels. For example, we could sub-divide the 2.4GHz range into 10 channels of 8.35 MHz each. Channels are identified by their center frequency, which is the frequency in the exact middle of the channel's frequency range. If we had a channel from 2.4 to 2.410 GHz, this channel would have:

- Center frequency of 2.405 GHz
- Bandwidth of 0.010 GHz or 10 MHz

Channel Advantages & Challenges

There are three primary types of interference:

- all band
- wideband (includes Spread Spectrum)
- narrowband

Often, interference will not take up the entire usable frequency range. Sub-dividing the range into channels gives the transmitter and receiver more options as to what frequencies they use. If interference occurs on one channel, wireless devices may use a different channel to avoid it. This principle is commonly used in cordless phones, which transmit on multiple channels in their frequency range simultaneously. The transmitter and receiver must agree on which channel to use.

The channel numbers in the transmitter and receiver must match. Various schemes, such as a mathematical algorithm, are used to determine the channel to use. In crowded frequency ranges, such as the 2.4GHz range, all channels may experience interference. Spread spectrum technology addresses this issue.

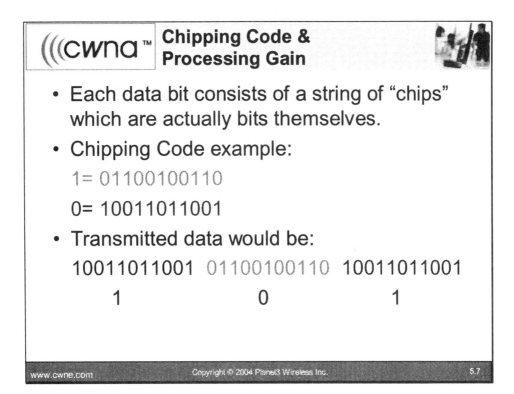

The higher the ratio of chips to data bits (processing gain), the better the resilience to narrowband interference. In DSSS, the chips are sent in parallel simultaneously on many frequencies across a wide frequency band. The sequence for a "1" and a "0" are compared to the received data bit (sequence of chips). If the number of chips is closer to the pattern of a "1", then the bit is deemed a "1". The same is true for a "0".

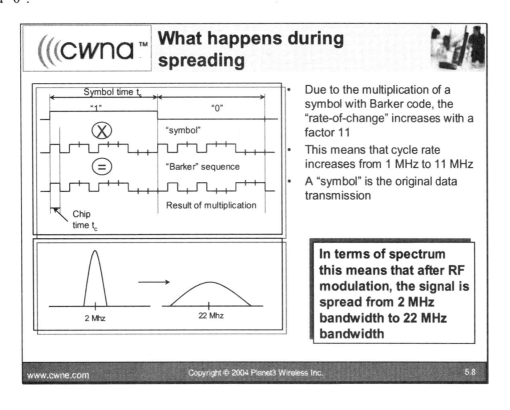

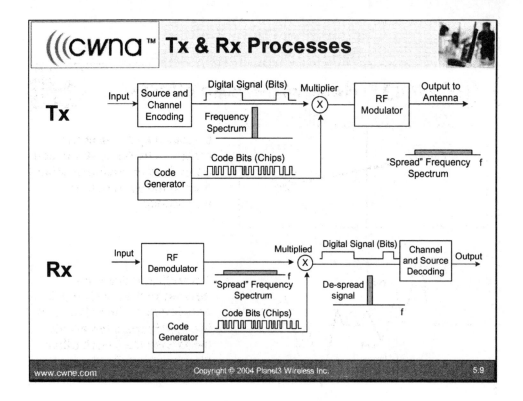

Spreading

The information signal is multiplied by a unique, high rate digital code which stretches (spreads) its bandwidth before transmission. Code bits are called "chips".

The sequencing method used in 802.11 is called *Barker Code*, although other coding types are now in use. At the receiver, the spread signal is multiplied again by a synchronized replica of the same code, and is "de-spread" and recovered. The outcome of the process is the original information.

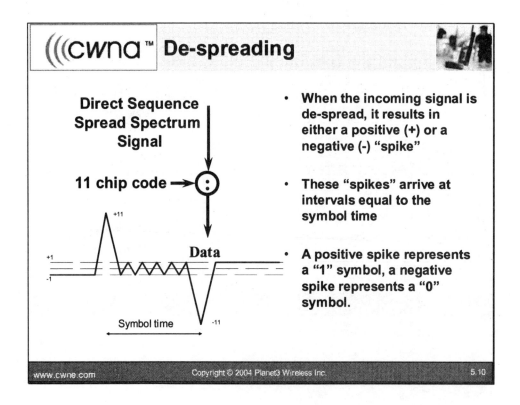

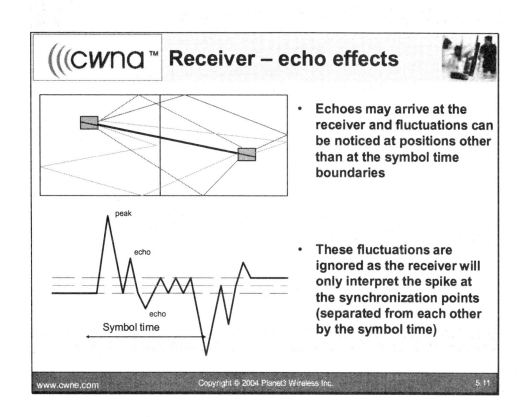

Chapter 6
Access Points, Service Sets, and the Distribution System

Objectives

Upon completion of this chapter you will be able to:

- Explain the functionality of access points and understand the 802.11 definition of an access point
- Describe the different service set types defined by the 802.11 standard
- Describe 802.11 distribution systems
- Understanding repeater functions and their advantages and disadvantages
- Explain the BSSID and ESSID and their use in a wireless LAN
- Understand various methods of mounting access point hardware

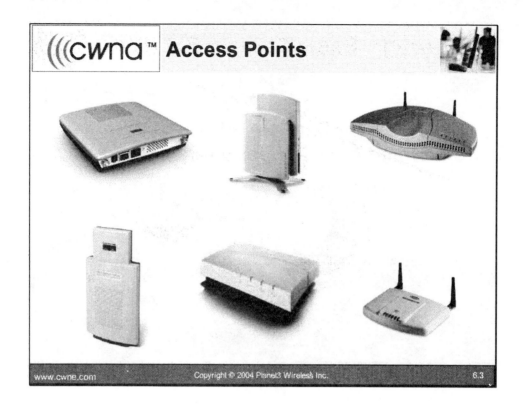

An access point is a device that contains an IEEE 802.11 conformant MAC and PHY interface to the wireless medium, providing access to a distribution system for associated stations.

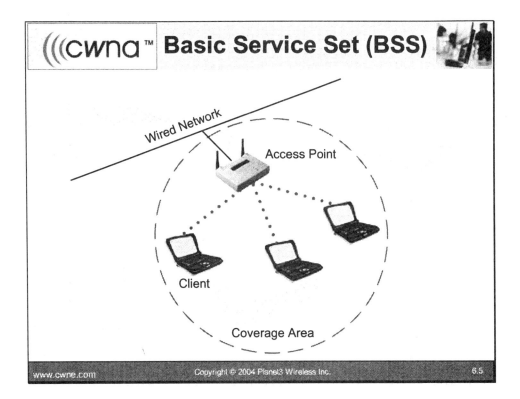

The Basic Service Set (BSS) is the basic building block of an 802.11 network. A BSS consists of an access point and a number of wireless stations. The access point may or may not be connected to a wired network infrastructure.

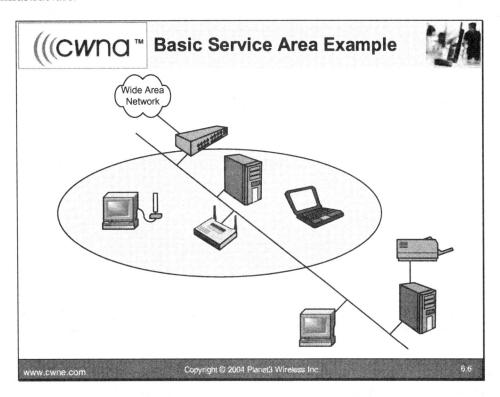

A BSA (Basic Service Area) refers to the conceptual area covered by a BSS.

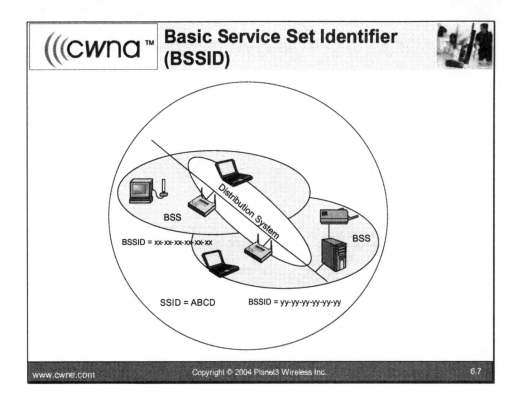

A BSS is identified by the 48 bit (6 octet) MAC address of its access point. This address is known as the BSSID (Basic Service Set Identification) and is used as the "cell identifier" and to identify the "old AP" during roaming.

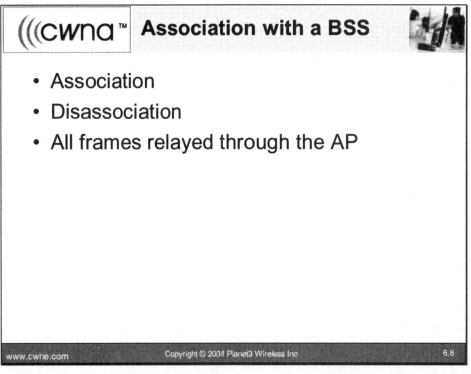

Stations dynamically become members of a BSS through a process called association and dynamically leave a BSS through a process called disassociation (covered later). Stations will attempt to associate with a BSS when they come into its BSA (coverage area). Stations will usually disassociate from a BSS when they leave the BSA of the BSS or power down. All traffic on a BSS must traverse the access point. Communication between two wireless clients means that wireless frames must be relayed through the access point.

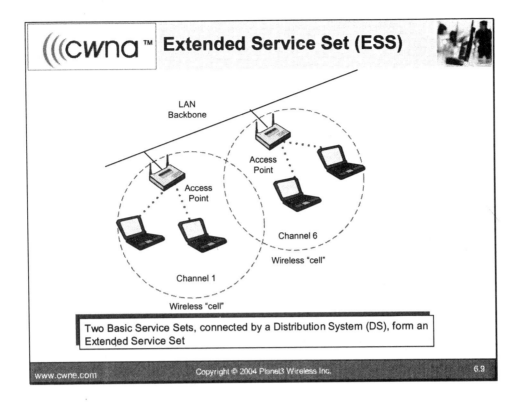

The ESS is identified by a variable-length string known as the "network name". This string is known as the ESSID (Extended Service Set ID), often as the SSID, and sometimes even the Net_ID. The ESSID can be a maximum of 32 ASCII characters and is case sensitive. All access points with the same ESSID form a single wireless network, or ESS. The method of configuring the ESSID varies depending on the manufacturer of the access point. The 802.11 specification avoids defining the nature of the distribution system. It could be any kind of network, as long as it provides certain services to the 802.11 communicators.

Wired Ethernet networks are commonly used as a distribution system. The distribution system and access points are responsible for enabling connectivity between stations in any BSS of the ESS. The access point must know which stations are associated with which BSS, and must handle getting messages from one access point to another.

Nothing is assumed by the 802.11 specification regarding the relative physical locations of the BSSs in the ESS.

- Basic Service Sets may be non-overlapping (disjointed)
- Basic Service Sets may partially overlap
- Basic Service Sets may be totally overlapping (co-located)

The SSID is NOT used for security purposes.

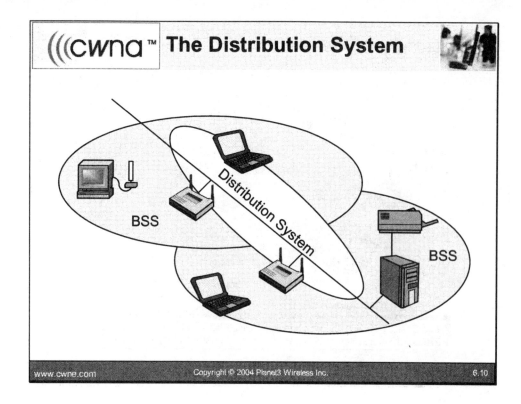

The DS is most often wired using a switched Ethernet connection due mostly to the use of end-span PoE switches. The DS can be anything, including wireless.

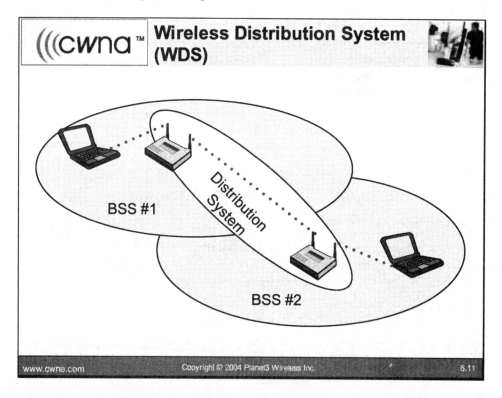

A wireless LAN link may be used as a distribution system. Distribution system radios must be on the same channel using the same technology. Using 802.11a/g access points in which one radio type is used for clients and another for the distribution system is preferable.

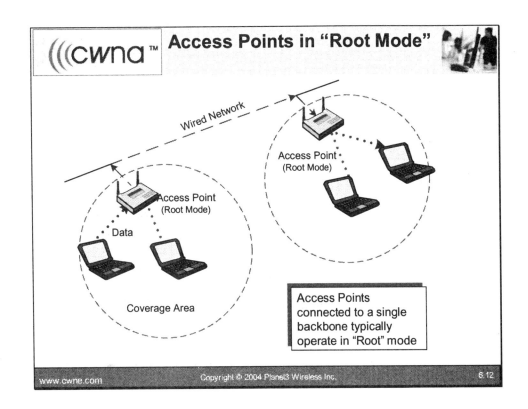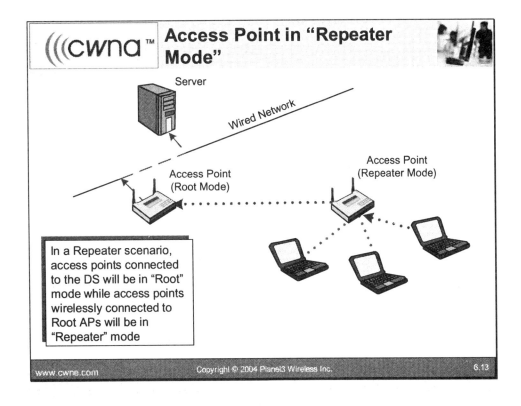

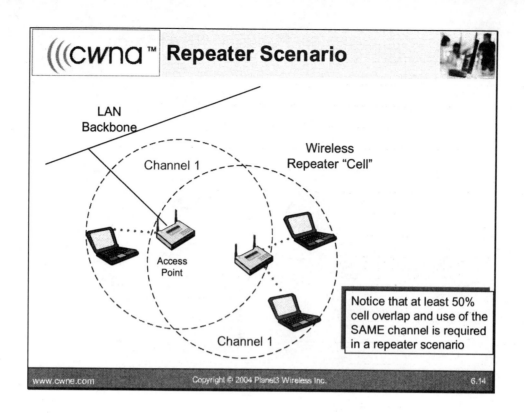

Chapter 7
Wireless Stations, Accessory Devices, and SOHO Networks

Objectives

Upon completion of this chapter you will be able to:

- Describe the various radio types available on the market today
- Understand the 802.11 definition of a Station (STA)
- Describe various converters and adapters for radio cards
- Describe SOHO Wireless Gateways (Routers) and their role in SOHO networks
- Understand the advantages of Ethernet and Powerline Networking in the home using wireless technology
- Describe various wireless client accessories and their use in SOHO wireless networks

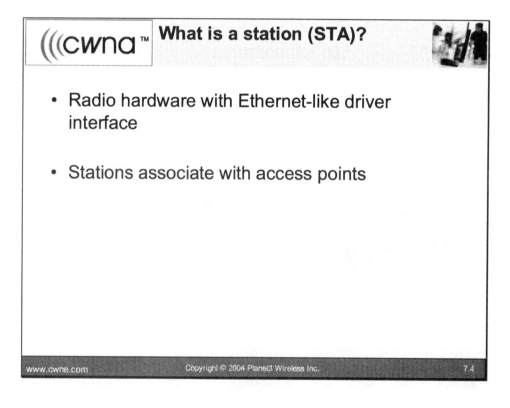

A wireless station (STA) is a device that contains IEEE 802.11 conformant MAC and PHY interface to the wireless medium, but does not provide access to a distribution system. Wireless stations have radio hardware with Ethernet-like driver interfaces, support virtually all protocol stacks, and associate with access points.

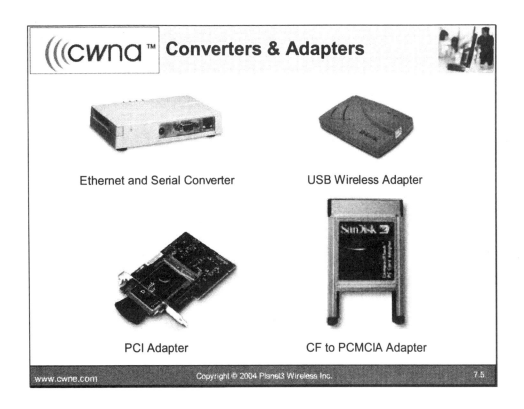
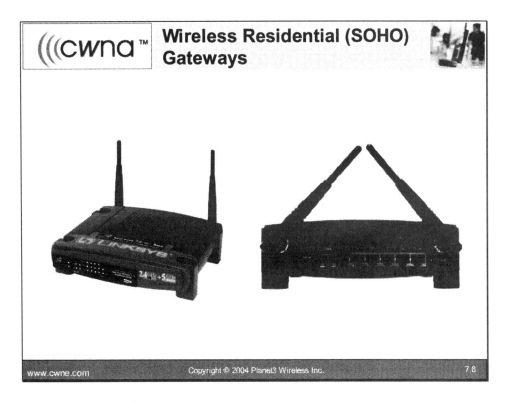

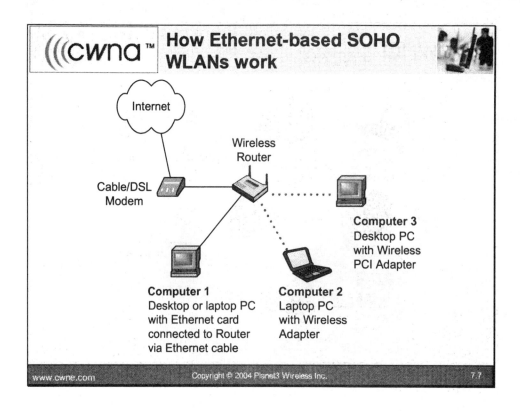

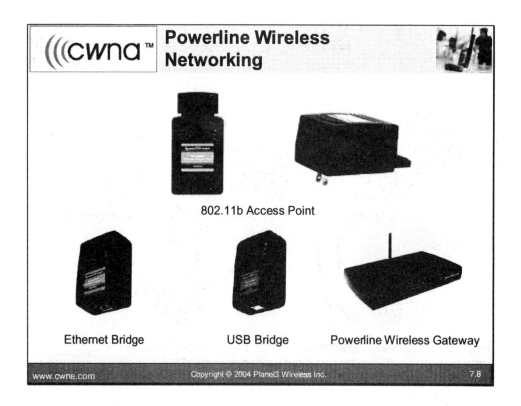

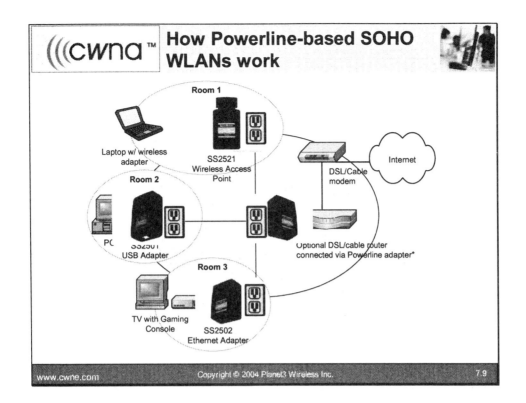

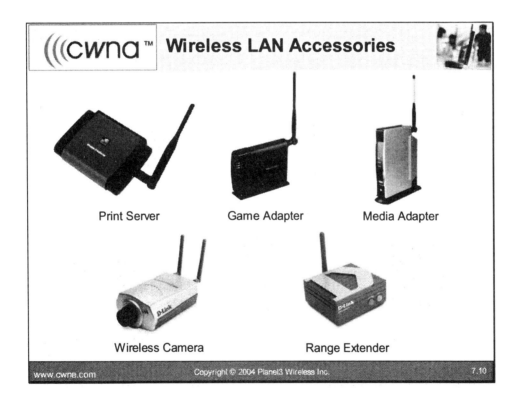

Chapter 8
Ad Hoc (IBSS) Networks

Objectives

Upon completion of this chapter you will be able to:

- Understand what components comprise an Ad Hoc wireless LAN
- Understand how routing is performed in an Ad Hoc network
- Understand how beacons and association works in Ad Hoc networks
- Understand how multiple Ad Hoc wireless LANs co-exist

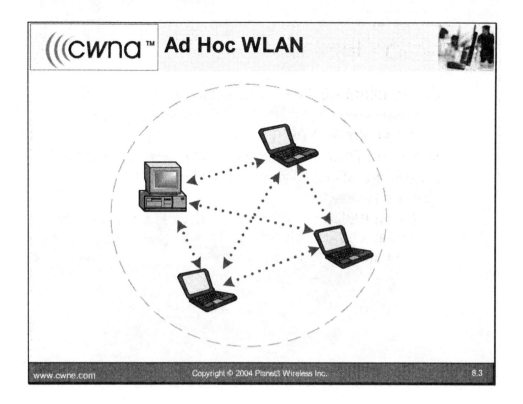

An ad hoc network is also known as an Independent Basic Service Set (IBSS). The simplest form of an 802.11 network is the IBSS (Independent Basic Service Set). An IBSS consists of a set of stand-alone wireless stations connected together as a group using the same network name (SSID). Traffic destined outside of the subnet formed by an Ad Hoc network must be routed by one of the peer nodes equipped with a second network interface.

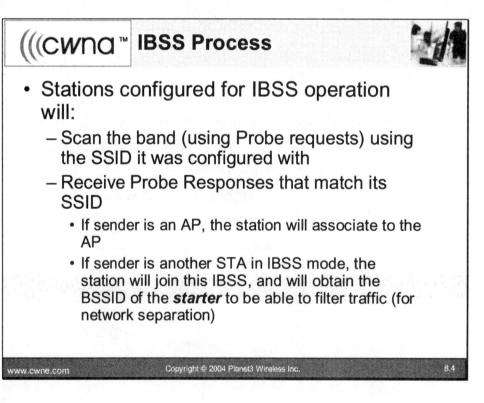

 IBSS Process

- Some client utilities allow the user to determine:
  - Whether the client will choose only an IBSS, only an AP, or any network
- When no Probe Responses are received with a matching SSID, the station will start its own IBSS network:
  - Set an BSSID (randomly generated, in MAC address format with local bit on)
  - Start sending Beacons

 IBSS Process

- All Stations in an IBSS network will participate in sending beacons
  - All stations start a random timer prior to the point in time when next Beacon is to be sent (random back off procedure)
  - First station whose random timer expires will send the next beacon

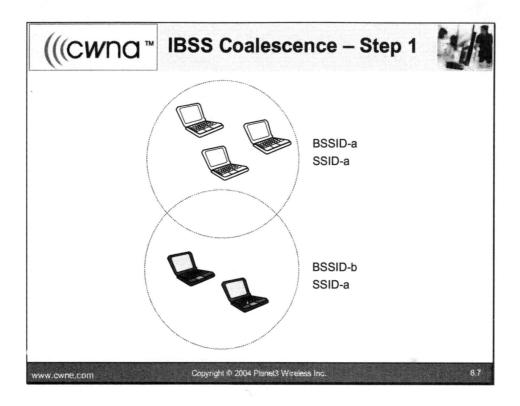

Different IBSS networks with the same SSID might exist, if cell members are out of each other's radio-range when the members of the network start up. In the illustration, two networks are shown with different BSSIDs: BSSID-a and BSSID-b. Both networks are configured with the same SSID (SSID-a).

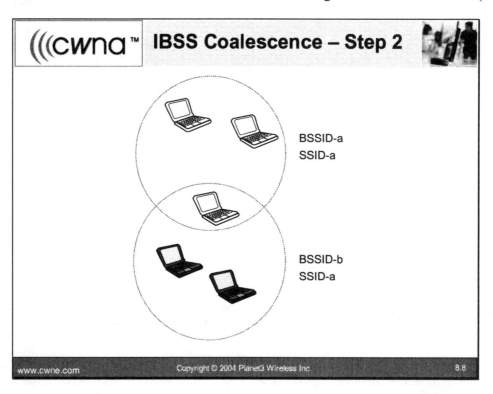

When a station moves, the station might get into radio-range of a neighboring cell (from BSSID-a to BSSID-b). The station will receive the beacons from the neighboring cell, and examine these for the SSID. It will find that these beacons contain the same SSID. Based on time-stamp information in all beacon messages received (from BSSID-a and BSSID-b), the station can join the other network.

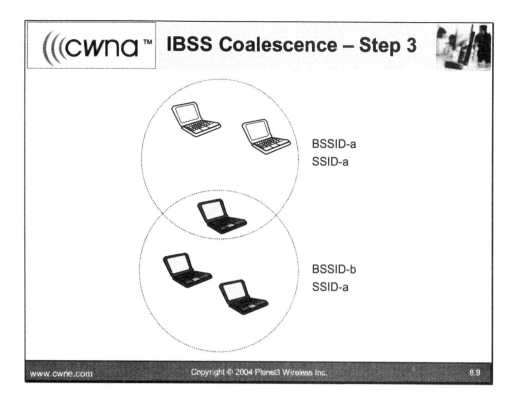

To join the network the station will obtain the BSSID from the frame header of the beacon message. Once joined, the station will participate in sending beacons and using the new BSSID (according to the coordination of the new cell). Other stations close to the last one that joined the new cell will be able to receive the beacons now as well.

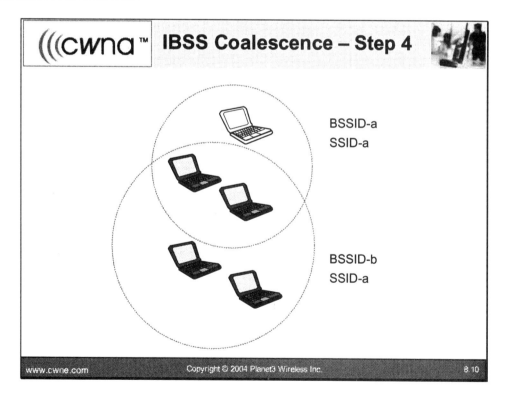

The process repeats itself and another station might add itself to the network. This process can continue until all stations might have joined the cell.

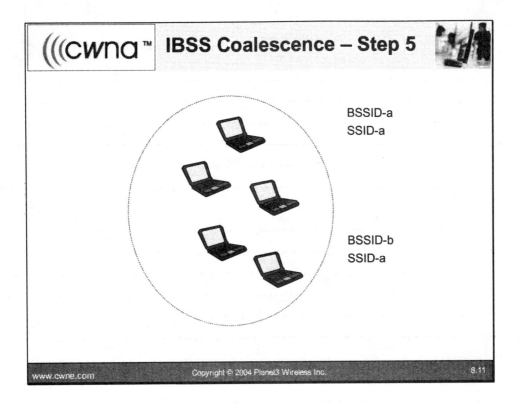

When the two cells have grown into one, it is known as "coalescence". When a station in this large cell starts missing beacons (e.g., no beacons have been received for 10 seconds), the station assumes that it is alone and then scans all channels. This station may find another access point or station that sends probe responses with a matching SSID, connect to that network, or start new IBSS (with new BSSID).

Chapter 9
Wireless Bridges and Workgroup Bridges

Objectives

Upon completion of this chapter you will be able to:

- Explain the functionality of wireless bridges and workgroup bridges
- Explain the different applications of wireless bridges and wireless workgroup bridges

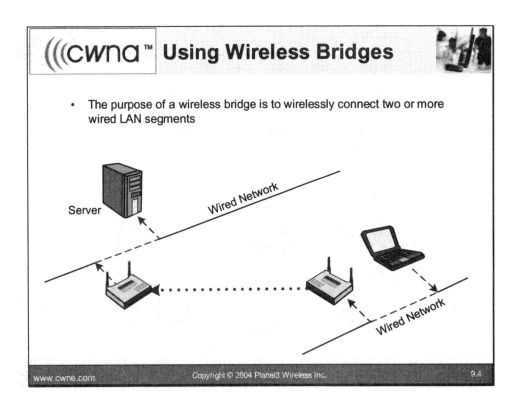

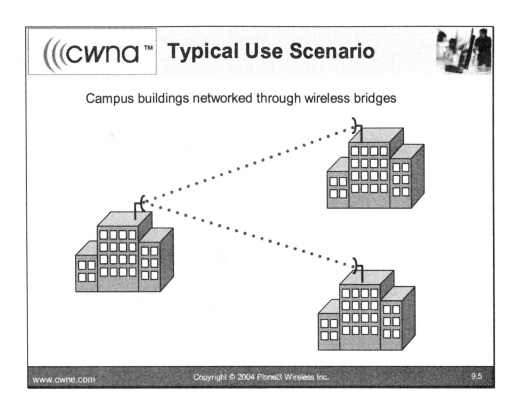

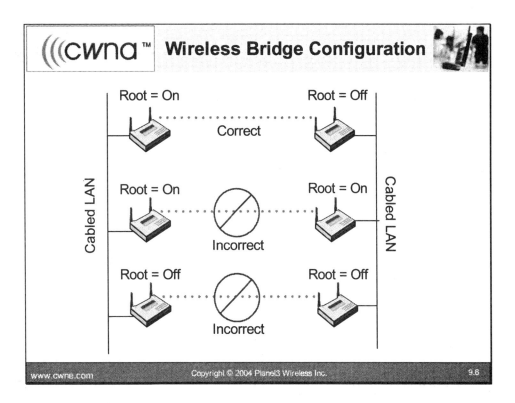

The bridge configured as Root=ON (Parent) communicates *only* with non-root bridges, clients, or repeaters, and will *not* communicate with other Root devices. Simultaneous client connectivity and bridging functions are vendor specific.

The bridge configured as Root=OFF (Child) communicates *only* with a Root or "Parent" bridge.

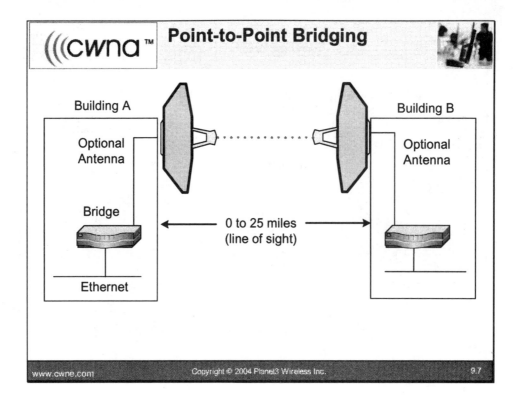

Aligning bridge links at long distances is very difficult. It is typically done on a roof or a tower using bridge alignment indicators and/or software tools.

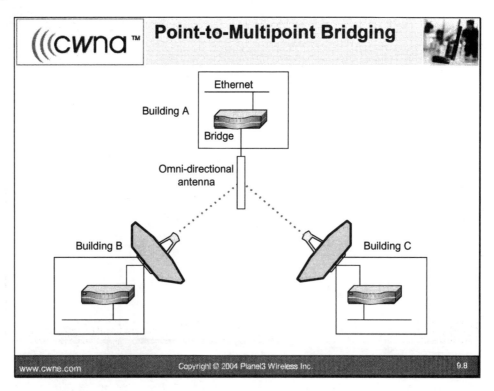

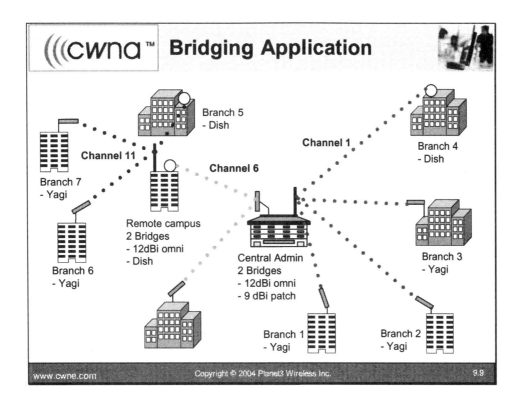

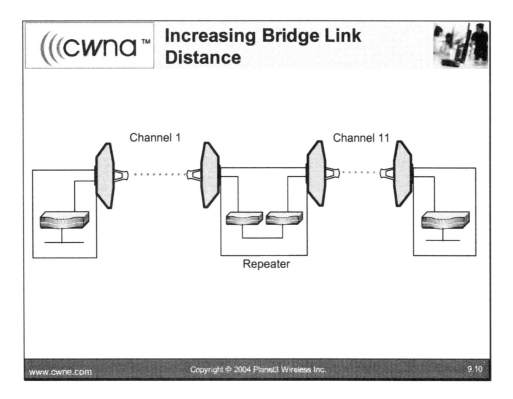

Use of repeater sites can allow longer distances to be spanned using wireless bridging.

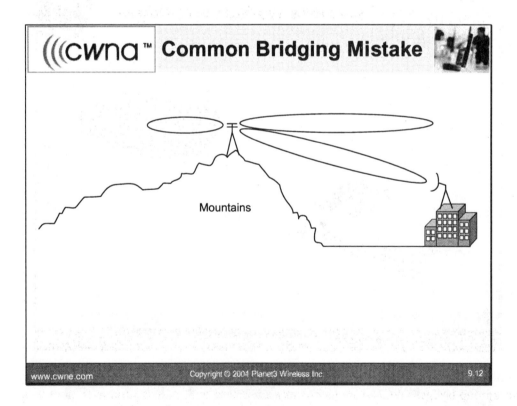

Bridge link calculations are not intuitive. Some vendors, like Cisco and Proxim, provide Excel spreadsheet tools for these calculations.

It is a common mistake to mount a high gain omni antenna too high. Keep in mind that RF connections are two-way.

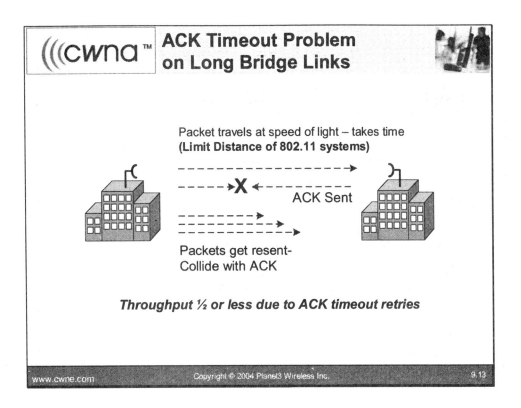

Though workgroup bridges (WGBs) are technically a wireless bridge, their purpose is more of a "collective wireless client" acting on behalf of multiple wired stations behind them.

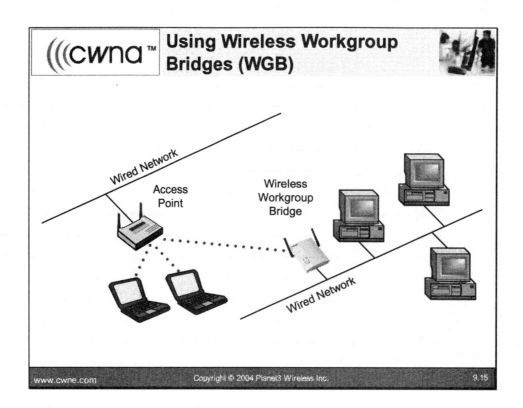

Chapter 10
Antennas, Cables, and Connectors

 Objectives

Upon completion of this chapter you will be able to:

- Explain differences between wireless LAN antenna types and how/when to use each type
- Explain FCC rules regarding antenna use with wireless LANs
- Explain what an isotropic radiator is and what antenna gain units are
- Describe the various wireless LAN connectors and cables and appropriate uses of each
- Explain the importance of polarization and antenna beamwidth

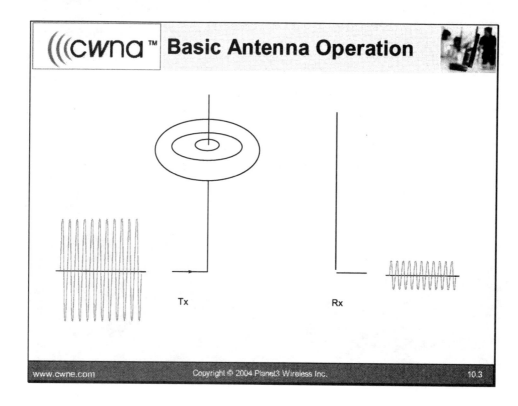

An electrical signal, fed into an antenna, can produce a standing wave, when the antenna matches the wavelength. Magnetic waves spread out at 90° from the electric standing wave on the antenna (wavefronts) as a result of the current flowing through the antenna. On the receiving antenna, the waves are converted to an electrical signal in the antenna.

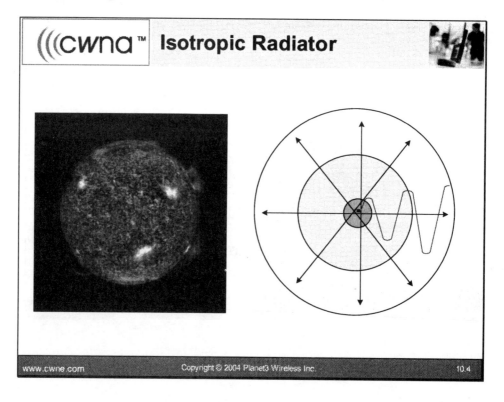

An isotropic radiator is an object that emits RF energy in a spherical radiation pattern – like our sun. The mathematical reference for an isotropic radiator is 0 dBi.

Antenna Gain Units

- Antenna gain is measured in either dBi or dBd.
  - dBi = decibels referenced against an isotropic radiator
  - dBd = decibels referenced against a dipole antenna
  - 0 dBd = 2.14 dBi

An Antenna's Physical Characteristics

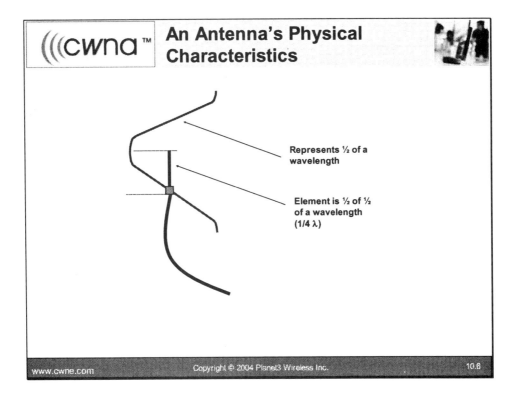

Antennas are made for a specific frequency based on dimensions of their elements.

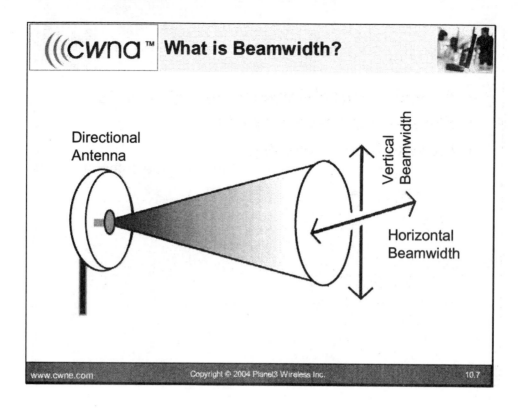

Coverage areas or "radiation patterns" are measured in degrees. These angles are referred to as beamwidths, and have a horizontal and vertical measurement. If the gain of an antenna goes up, the coverage area or angle goes down.

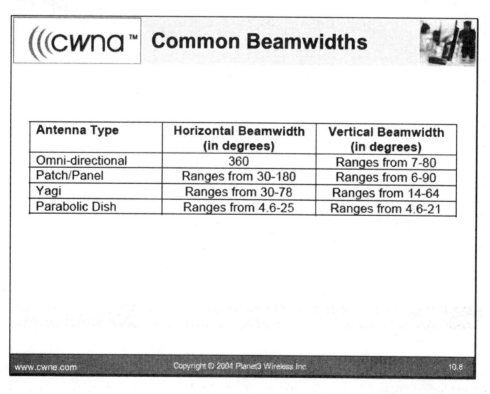

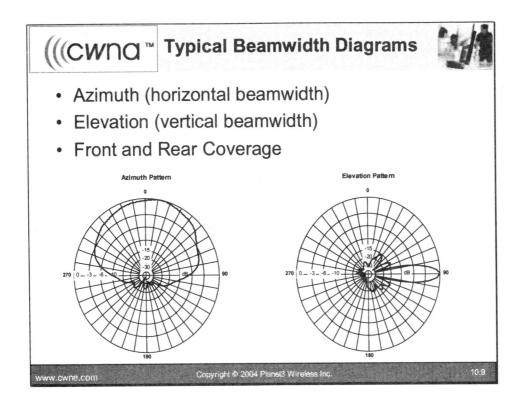

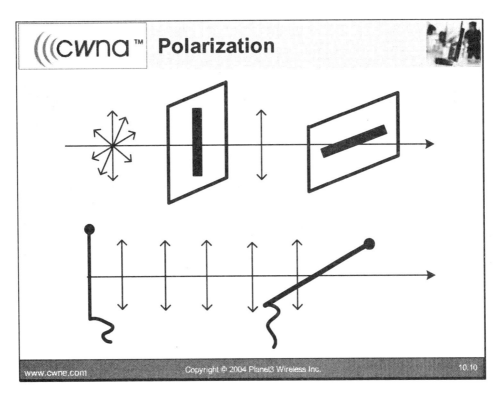

Consider two pieces of cardboard with long, thin slits in each one. Orient one slit vertically and the other horizontally and place them across a room. Shine a flashlight from behind one piece of cardboard. What will a person standing behind the other piece of cardboard see? Only a dot. This is representative of improper polarization matching between systems. Stations and access points must have antennas that are both polarized and positioned the same way to achieve good coverage.

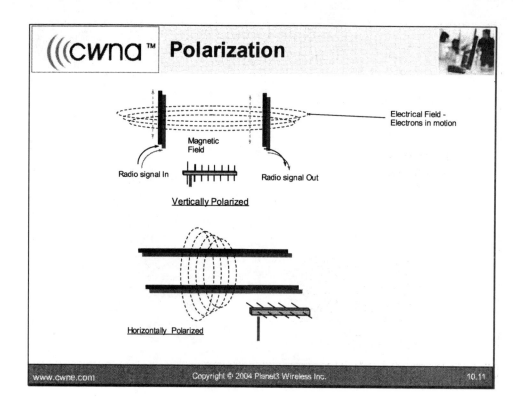

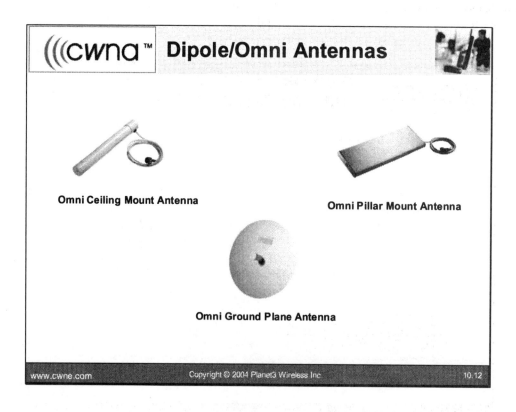

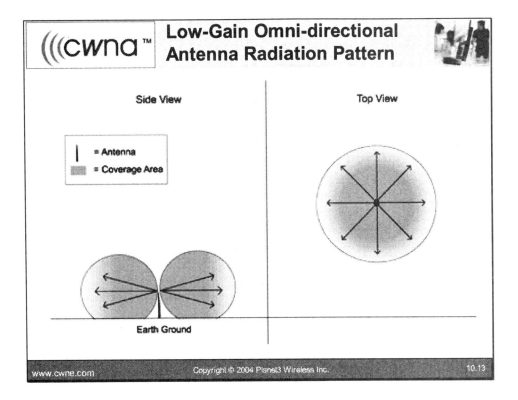

"Omni-directional" is a little misleading because all antennas exhibit some gain. This gain causes the radiation lobes to be squeezed and not perfectly round.

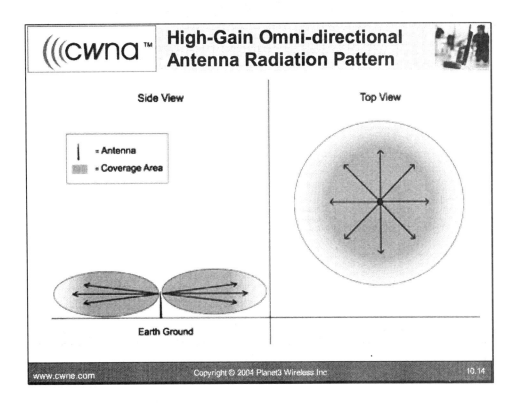

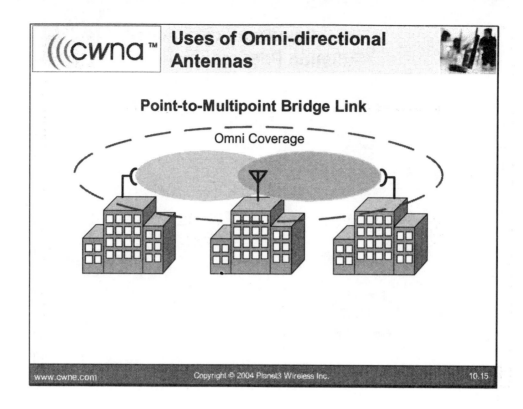
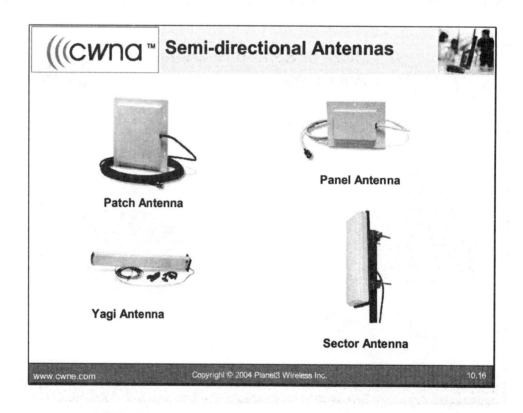

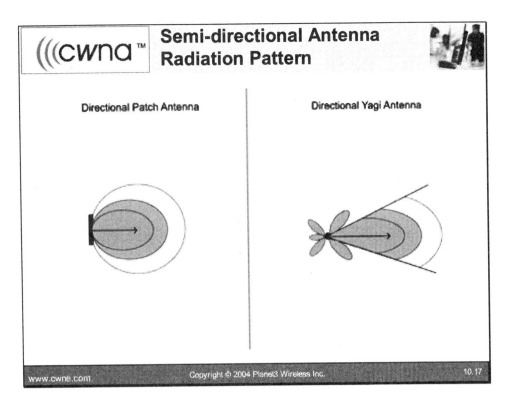

Both the Elevation and Azimuth should be taken into account when selecting a semi-directional antenna.

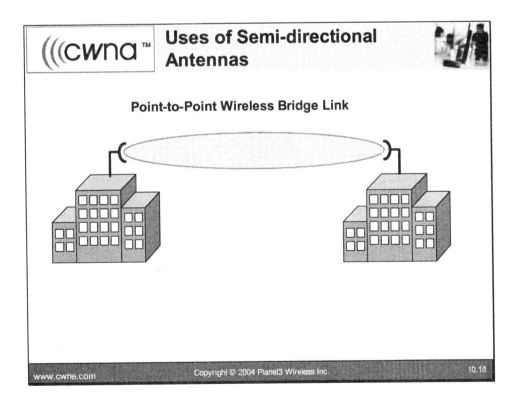

Corridors, hallways, alleys, long/narrow offices, and other similar situations are perfect indoor uses of semi-directional antennas.

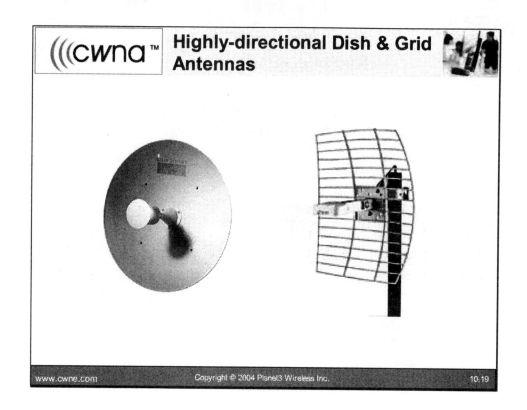

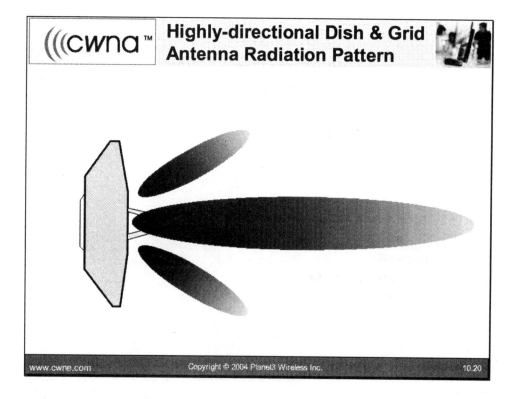

Most of the antenna's radiated energy is in the main lobe.

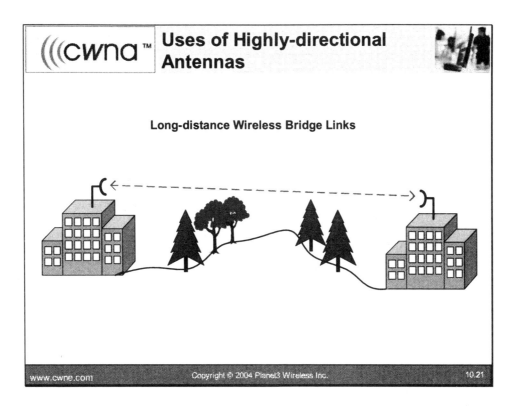

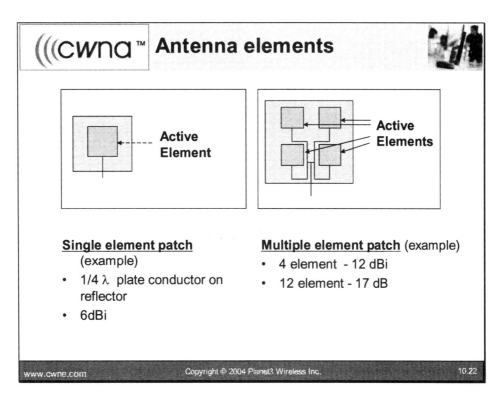

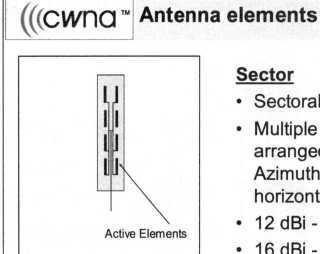

Sector
- Sectoral Dipole Array
- Multiple dipoles arranged to give large Azimuth pattern for horizontal coverage
- 12 dBi - 120°
- 16 dBi - 90°

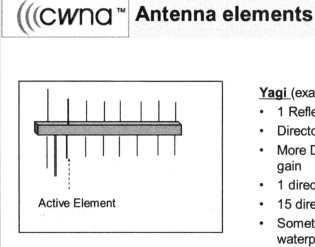

Yagi (example)
- 1 Reflector
- Directors
- More Directors - Higher gain
- 1 director = 8dBi
- 15 directors = 14 dBi
- Sometimes hidden in non-waterproof enclosure

 **Antenna elements**

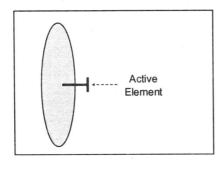

- **Parabolic** (example)
- Parabolic reflector focuses signal
- Larger Reflector = more gain
- 25 cm - 15dBi
- 1 m X 50 cm - 24 dBi
- 1 m full - 27 dBi
- 2m full - 31 dBi
- 3m full - 37 dBi

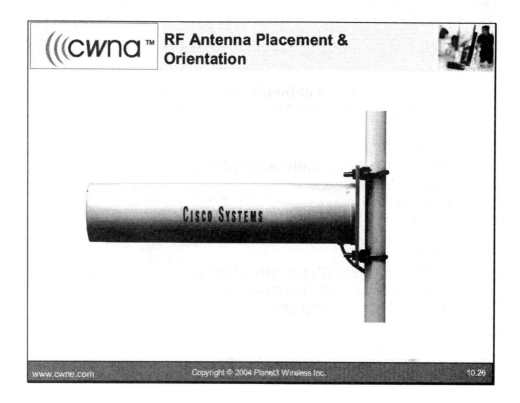

Placement & Orientation for Best Coverage

It is important that antennas be placed at the best possible location to ensure maximum connectivity. A site survey will determine what the best locations are. When placing and orienting an antenna, consider these questions:

- Where are the transmitters and receivers (clients and APs)?
- Where is the signal going to go (based on antenna type)?
- What is going to interfere with that signal, and how much?
- Is the Fresnel Zone clear? (discussed later)

Location of transmitters has a much larger effect on transmission quality and data rate in wireless communication than it does in wired communication. Physical obstacles may interfere with RF signals. Electromagnetic interference may be stronger in some areas. An antenna's ability to transmit around corners or down hallways is variable.

Choosing an antenna

When choosing an antenna, evaluation of the following points is highly recommended:

1. Gain
2. Beamwidths (horizontal and vertical)
3. Rear lobe coverage
4. Polarization
5. Cost
6. Intended use
7. Manufacturer
8. Impedance, VSWR, and other electrical characteristics
9. Attached cable and connector types
10. Available mounting gear

Antenna Safety

- Read factory manuals
- Keep your distance from transmissions
- Avoid metal obstructions
- Consider professional installers
- Avoid power lines
- Use grounding rods

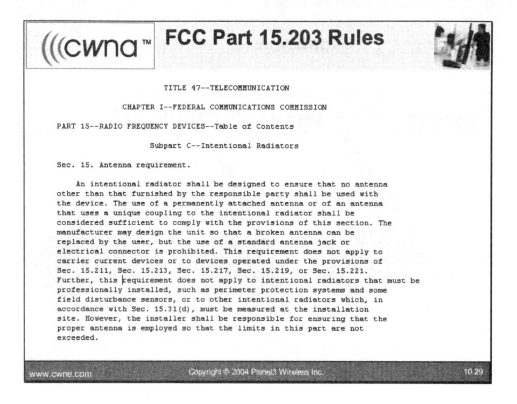

The FCC requires that ALL antennas sold by a spread spectrum vendor be certified for use with the radio for which they are intended. FCC requires that antenna connectors (couplings) be unique to the manufacturer.

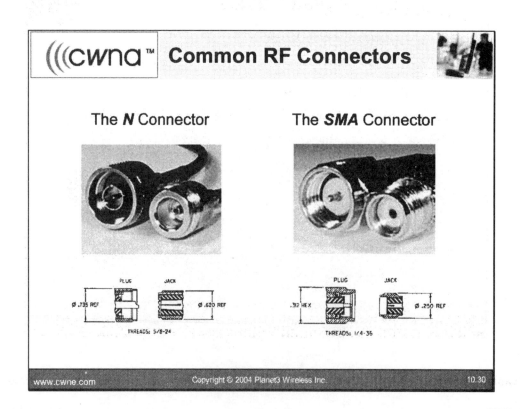

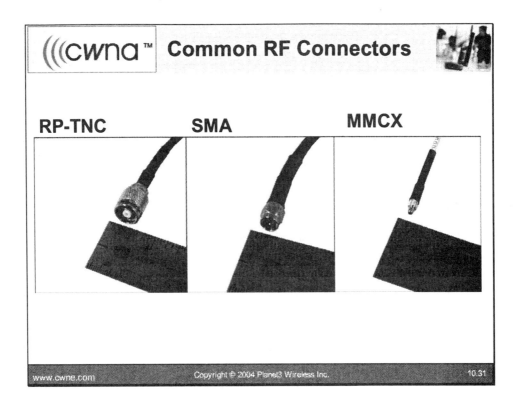

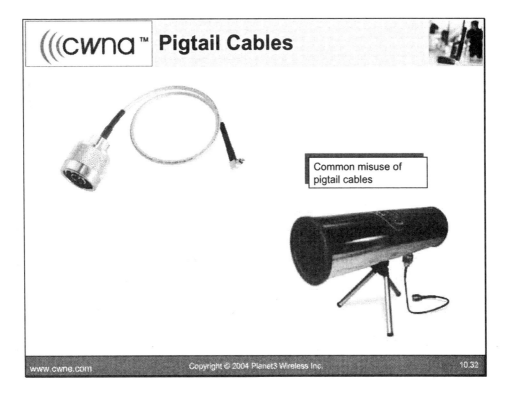

Pigtail cables are used to connect an antenna to a cable or device which uses a different connector.

WARNING: Remember the FCC rules on certified systems. Pigtail cables are acceptable if they are certified as part of the system, but arbitrarily adding a pigtail connector to an existing system just so that a device can have a better antenna is against FCC rules.

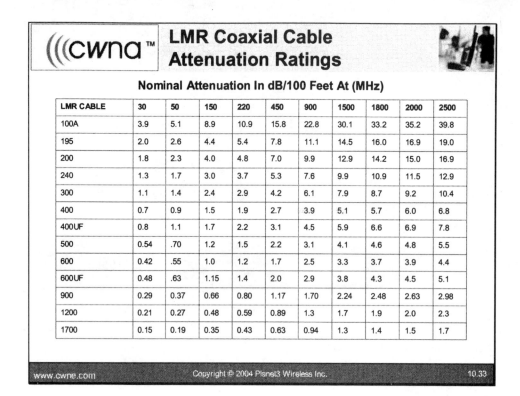

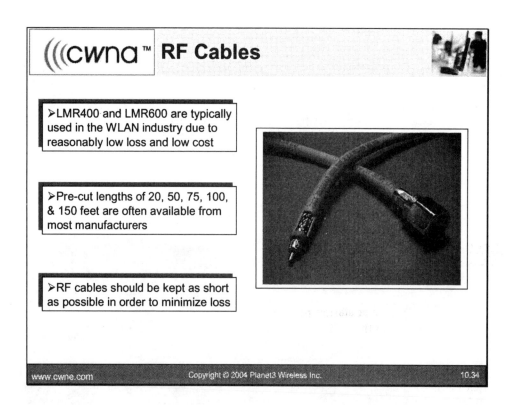

Pre-cut, professionally-made RF cables (typically provided by wireless LAN equipment manufacturers) are highly preferable to home-made cables. Low-loss cables are always preferable, but are significantly more expensive than lesser quality cables. Equipment manufacturer cables are always certified as part of the manufacturer's certified systems.

Chapter 11
Amplifiers, Attenuators, Splitters, and Lightning Arrestors

Objectives

Upon completion of this chapter you will be able to:

- Explain FCC rules regarding use of amplifiers
- Explain the functionality, different types, appropriate applications, and common specifications of the following wireless LAN accessories:
  - Amplifier
  - DC power injector
  - Attenuator
  - RF splitter
  - Lightning arrestor
  - Frequency converter

FCC Rules Regarding Amplifiers

Federal Communications Commission § 15.205

requirement does not apply to intentional radiators that must be professionally installed, such as perimeter protection systems and some field disturbance sensors, or to other intentional radiators which, in accordance with §15.31(d), must be measured at the installation site. However, the installer shall be responsible for ensuring that the proper antenna is employed so that the limits in this part are not exceeded.

[54 FR 17714, Apr. 25, 1989, as amended at 55 FR 28762, July 13, 1990]

§ 15.204 External radio frequency power amplifiers and antenna modifications.

(a) Except as otherwise described in paragraph (b) of this section, no person shall use, manufacture, sell or lease, offer for sale or lease (including advertising for sale or lease), or import, ship, or distribute for the purpose of selling or leasing, any external radio frequency power amplifier or amplifier kit intended for use with a Part 15 intentional radiator.

(b) A transmission system consisting of an intentional radiator, an external radio frequency power amplifier, and an antenna, may be authorized, marketed and used under this part. However, when a transmission system is authorized as a system, it must always be marketed as a complete system and must always be used in the configuration in which it was authorized. An external radio frequency power amplifier shall be marketed only in the system configuration with which the amplifier is authorized and shall not be marketed as a separate product.

(c) Only the antenna with which an intentional radiator is authorized may be used with the intentional radiator.

[62 FR 26242, May 13, 1997]

§ 15.205 Restricted bands of operation.

(a) Except as shown in paragraph (d) of this section, only spurious emissions are permitted in any of the frequency bands listed below:

Manufacturers of RF amplifiers must market them as part of a certified system. When using an amplifier, you must always make sure it is used as part of a certified system.

The European Telecommunication Standardization Institute (ETSI) has developed standards that have been adopted by many European countries. Under ETSI standards, the power output and EIRP are much different than in the US, as they specify a maximum EIRP as 20dBm. Since EIRP includes antenna gain, this output power limit reduces the number of antennas that can be used with a transmitter. To use a high gain antenna, the transmitter power must be reduced drastically. This limitation obviously reduces the distance at which an outdoor link can operate and since the ETSI regulation has such a low EIRP, the use of amplifiers are typically not permitted in any ETSI system.

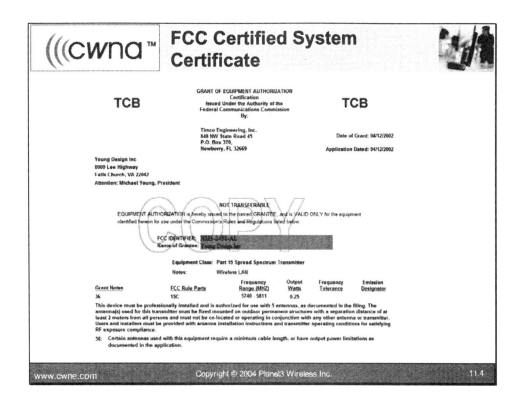

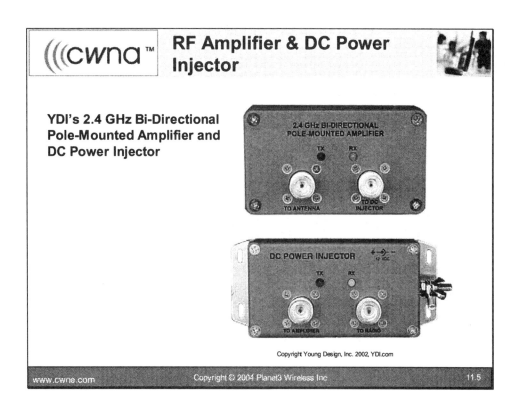

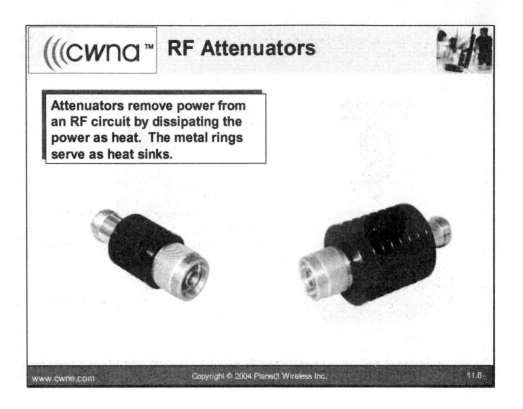

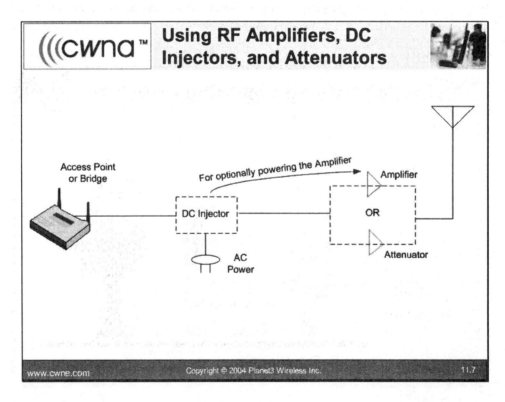

Amplifiers should usually be placed close to the antenna so that small signals being received by the antenna can be amplified before traversing an RF cable that will introduce even more loss.

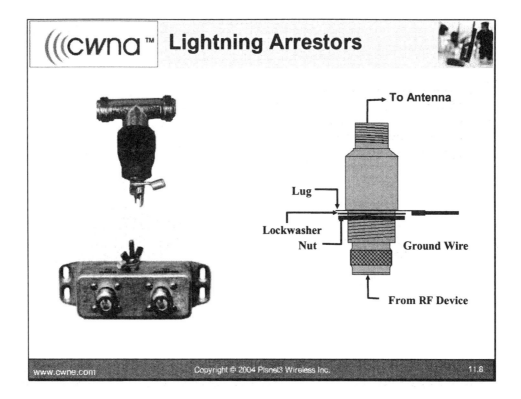

There are many different types of lightning arrestors. Some use gas cylinders, some use the physical properties of the conductor, and still others use capacitors. It is important to read the specs and operational information before purchases.

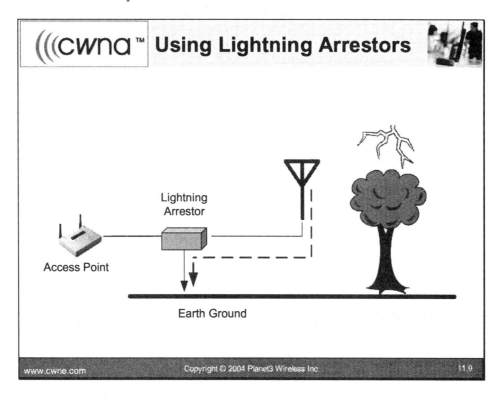

As objects near an antenna are struck by lightning, an electric field is created around those objects. Electrical current from that electric field is induced into the wireless LAN antenna causing damage to most systems.

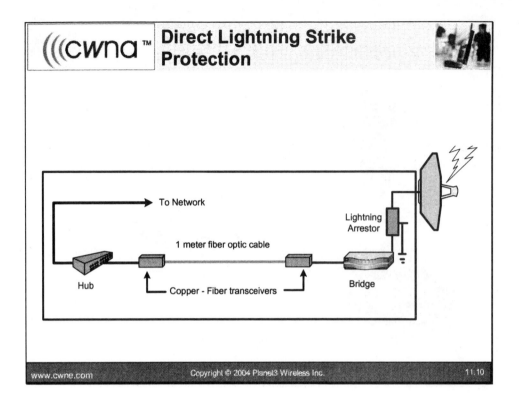

Though you would lose your wireless LAN bridge, your system can be saved from a direct lightning strike by introducing a fiber data connection long enough that lightning could not jump across.

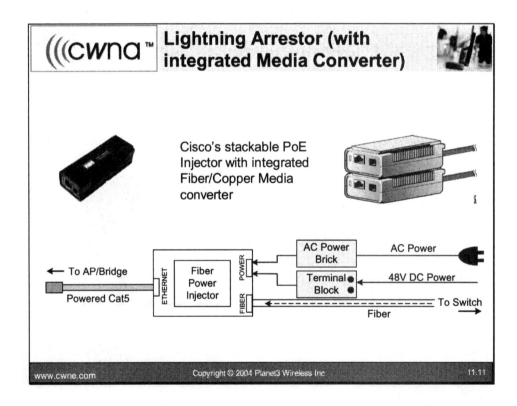

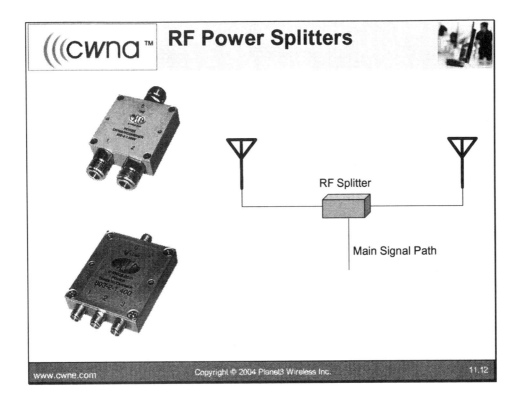

Splitters come with a variety of port densities, connector types, and impedance. Specifications for insertion loss, VSWR, frequency response, and other parameters vary greatly between models.

Splitters introduce significant loss into an RF circuit. The System Operating Margin (SOM) should be calculated to include the splitter loss before implementation.

Splitters should not be used unless the network designer can specifically justify their use. A splitter may be used along a long, linear coverage path with directional antennas pointed in opposite directions to cover sidewalks, highways, along rivers. Splitters may be used with sector antennas to cover a large circular area using multiple sectors for a WISP.

Chapter 12
Fresnel Zones, Free Space Path Loss, Interference, Fading, and Multipath

Objectives

Upon completion of this chapter you will be able to:

- Define and explain the importance of:
  - Fresnel Zones
  - Earth bulge
  - Antenna height calculations
  - Free Space Path Loss
  - Common sources of RF interference
  - Multipath fading
  - Signal distortion
  - Antenna diversity

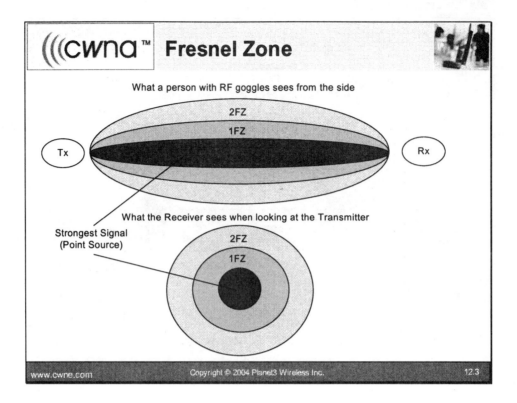

Antennas have the advantage of seeing "phases," which is different from how a human would see light. Fresnel Zones are in-phase and out-of-phase areas (zones) around the point source. The First Fresnel Zone is loosely defined as the area between the transmission point source and the end of the first in-phase zone surrounding the point source. Picture in your mind a dot with concentric circles around it. The area between the center dot (called the point source) and outside of the first ring is considered the First Fresnel Zone (1FZ). Do not block the 1FZ because it is in-phase (additive) with the main transmission signal (the dot).

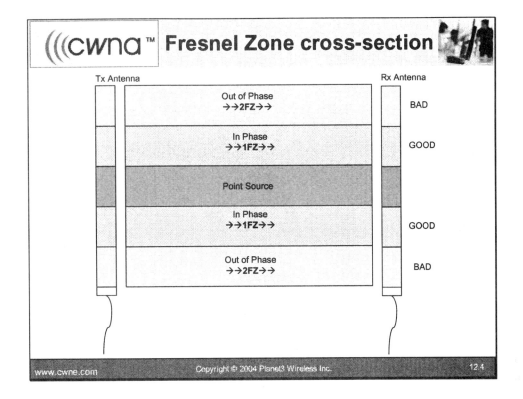

The first zone adds to the main signal amplitude (because it is in-phase), and the second zone (2FZ) subtracts from the main signal amplitude. Zones alternate between in- and out-of-phase thereafter as well. It is acceptable, even advantageous, to block the 2FZ, but it is not typical to attempt to block it on purpose.

Imagine standing on a tower at night looking at another tower which had a simple flashlight bulb shining from it. This situation is representative of an omni antenna propagating RF waves. What would you see? You would see a dim dot of light (the point source). If you had binoculars (representative of a high-gain receiving antenna) what would you see? You would see a bright dot of light. If someone took the flashlight reflector out of the flashlight and put it on the other side of the light so that all of the light was reflected toward you, what would you see without the binoculars? You would see a bright dot. What would you see with the binoculars? You would see a really bright dot. The point is that the receiver only sees a point source with in- and out-of-phase rings around it. Antenna gain and beam width changes only affect perceived intensity (power) of the source.

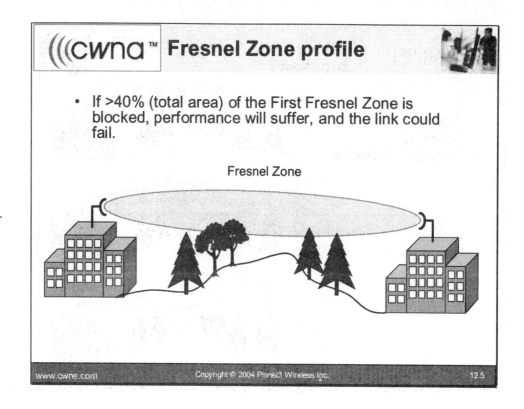

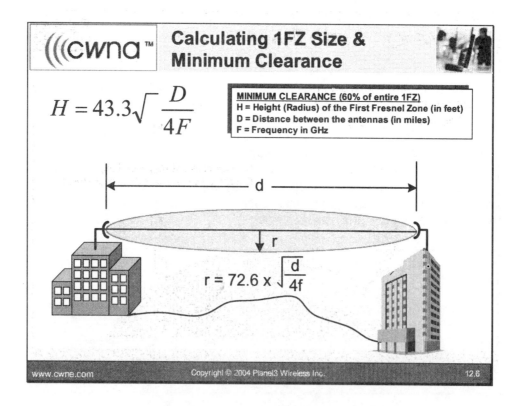

Because of the shape of the First Fresnel Zone, what appears to be a clear line-of-sight path may not be completely clear. As long as 60% of the First Fresnel Zone is clear of obstructions, the link behaves essentially the same as a clear free-space path.

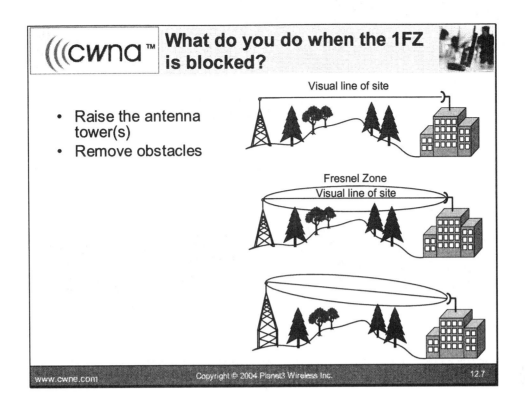

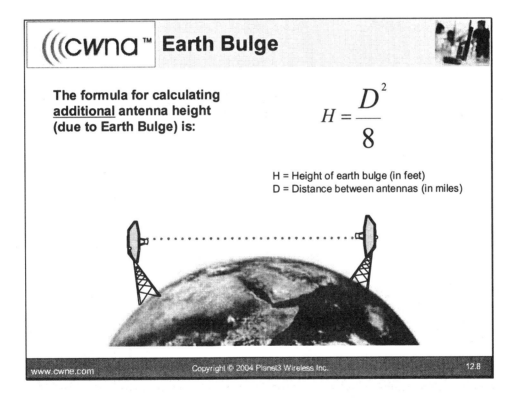

For bridge links longer than 7 miles, Earth bulge becomes a factor when determining Fresnel Zone clearance.

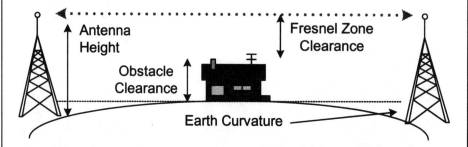

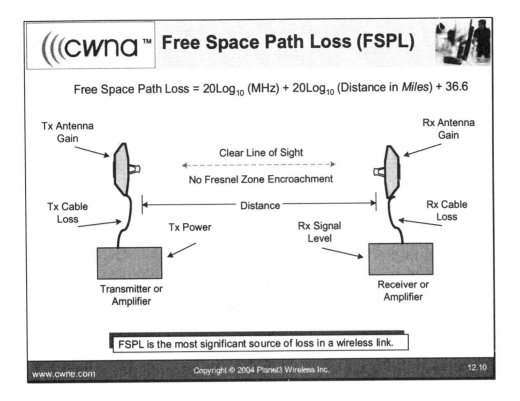

Free space path loss (FSPL) is a reduction in the amplitude of a signal due to natural expansion of the wave front. As an illustration of this concept, consider a simple balloon. As the balloon is filled with air, the surface of the balloon gets thinner and thinner.

Three main factors dictate how much of a transmitted signal is received by the receiver:

1. The distance between transmitter and receiver
2. How focused the transmitted signal is in the direction of the receiver
3. How much of the transmitted signal the receiver can grab (size of the receiving aperture)

Three main factors dictate how useful the received signal will be:

1. The power of the transmitted signal
2. The sensitivity of the receiver
3. How much interference was present (other signals, blockage, distortion, multipath, etc.)

FSPL is the mathematically calculated power loss that would occur if a signal were transmitted between two antennas that were separated by a vacuum. FSPL disregards any attenuation caused by the signal passing through air, walls, etc. FSPL is a useful way of getting a perspective on RF attenuation levels. Actual attenuation will always be greater than the mathematically calculated FSPL.

For a 2.4 GHz signal, the FSPL of various distances is:

- 100 meters = -80 dB
- 320 meters = -90 dB

If a signal is transmitted at 1 mW, and the receiver sensitivity of the card is about –90 dBm, then about 300 meters is the maximum range.

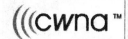

Free Space Path Loss Calculation

$L_p = 36.6 + (20 \log_{10} F) + (20 \log_{10} D)$

…where

> L_p = free-space path loss between antennas (in dB)
> F = frequency in MHz
> D = path length in *miles*

…or

$L_p = 32.4 + (20 \log_{10} F) + (20 \log_{10} D)$

…where

> L_p = free-space path loss between antennas (in dB)
> F = frequency in MHz
> D = path length in *kilometers*

A signal degrades as it moves through space. The longer the path, the more loss it experiences. Thus, free-space path loss is a factor in calculating the link viability. Free-space path loss is easily calculated for miles or kilometers using one of these formulas.

Signal power is closely related to interference. Weaker signals are more likely to receive interference and less likely to cause it. More powerful signals are more likely to cause interference and less likely to receive it.

Ambient electromagnetic energy ("noise") can interfere with RF signal reception. Energy that is similar to the RF signal is more likely to interfere with the signal. A 900MHz phone is less likely to interfere with a 2.4GHz 802.11 LAN. A 2.4GHz phone is more likely to interfere with a 2.4GHz 802.11 LAN.

Other potential sources of interference include:

- Non-ISM systems (Electrical Devices)
- Elevator motors
- Overhead cranes with heavy spiking electric motors
- Welding equipment
- Power lines
- Electrical railroad track
- Power stations
- Microwave Ovens
- Fluorescent Bulbs
- Obstructions
- Walls, windows, or duct work that contain metal or wire mesh
- Metal desks & cabinets
- Metal structures, buildings, etc
- Moving objects: aircraft, cranes, vehicles

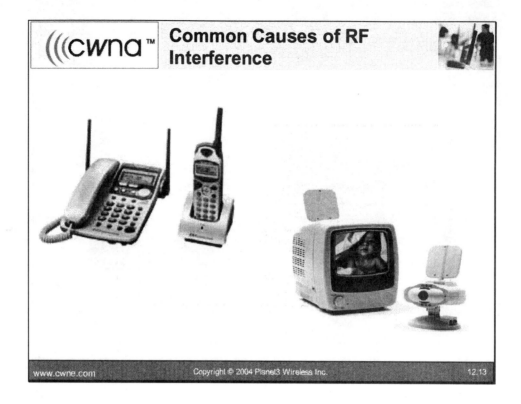

Wireless LANs (FHSS, DSSS, OFDM) typically use low power if adhering to regulations. Such systems include:

- 2.4 GHz Baby Monitors
- 2.4 GHz spread spectrum phones
- Bluetooth-enabled devices

Bluetooth is a wireless technology similar to 802.11 that uses the same 2.4GHz frequencies but has a shorter range (tens of meters vs. Hundreds), much lower data rate (721 Kbps vs. up to 11 Mbps), and uses smaller form-factor transmitters.

Steps are being taken by the relevant standards bodies to ensure minimal interference between Bluetooth and 802.11. Interference can occur up to about 50 feet. Multipath interference occurs when signal reflects off of objects and interferes with itself.

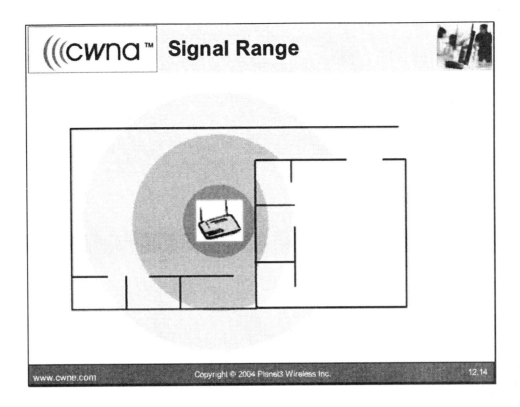

Range is the effective distance at which the signal can be reliably received. The more attenuation and interference a signal encounters, the lower the range of the signal. 802.11 cards have maximum ranges varying from 100 meters (indoors) to between 300 and 800 meters (outdoors). Through walls and ceilings, this range can be reduced to around 20 or 30 meters.

High-gain antennas can transmit 802.11 signals much further. Higher ranges are possible with more expensive technology. Line-of-sight requirements limit distance as obstacles impinge upon the line of sight between transmitter and receiver, range decreases. FCC power regulations limit distance. The FCC requires special licensing for RF signals above certain power levels, even in the 2.4 GHz range.

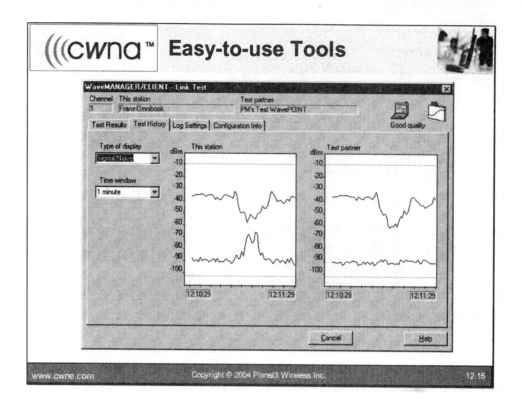

Some vendors include noise-measurement software with their site surveying tool set included with the client software. You can avoid the interfering frequencies by configuring access points to utilize channels that do not use the same frequencies as the interference source.

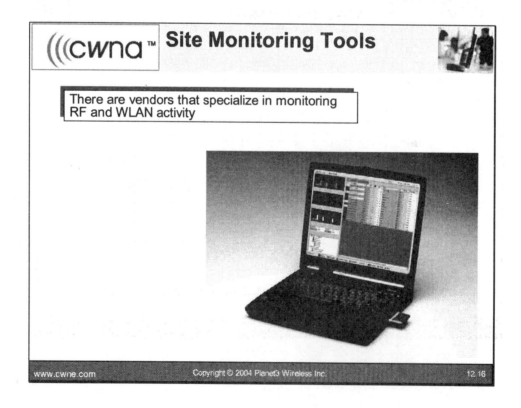

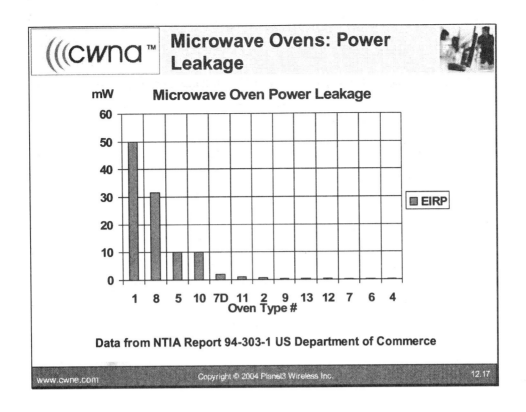
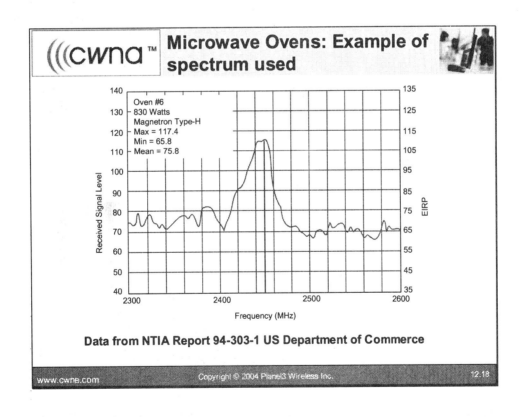

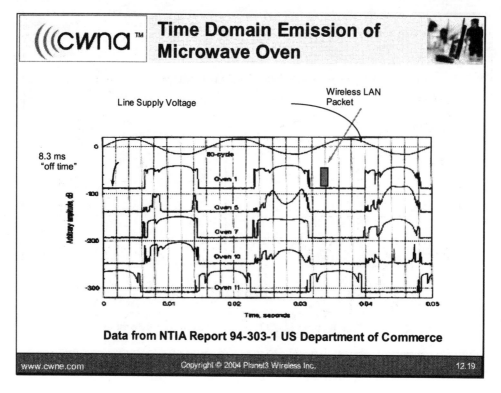

Microwave ovens use what is known as an on/off cycle. The "Off" cycle can be used to get wireless transmissions through. Depending on the power cycle (50 or 60 Hz), the "off time" equals to 10 or 8.3 msec. Some manufacturer's have a "Microwave Oven Present" type setting – sometimes called "Interference Robustness".

Transmitting a max size packet (1500 bytes) takes approximately:

- 12.5 msec @1 Mbps
- 6.2 msec @2 Mbps
- 2.3 msec @5.5 Mbps
- 1.1 msec @11 Mbps

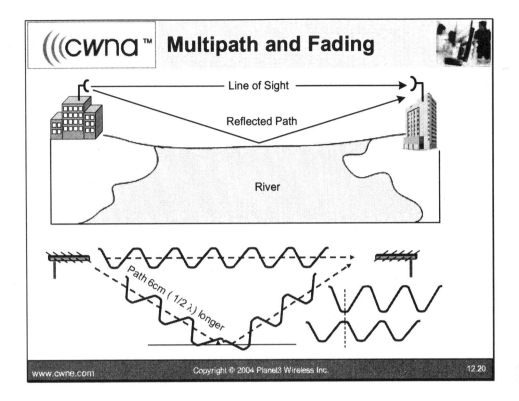

Rayleigh Fading (aka. Multipath Fading) is an increase or decrease in amplitude of the main RF transmission signal due to reflected signals (in-phase and out-of-phase) being combined with the signal along the transmission path. Terms such as 'upfade' and 'downfade' are commonly used to denote amplitude shifts (up or down) due to multipath fading.

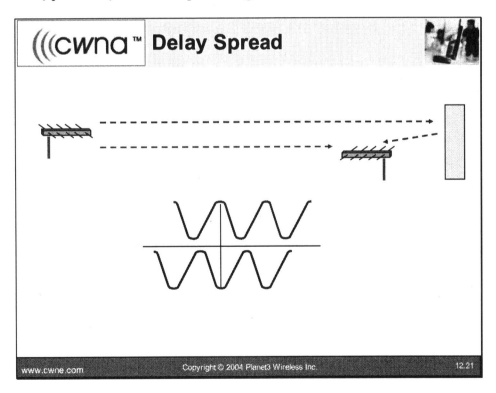

Changes as small as 6 cm in an antenna's position can mean the difference between in-phase and out-of-phase. The time differential between the arrival of the main signal and the reflected signal is called "delay spread". Delay spread is usually less than 4 nanoseconds.

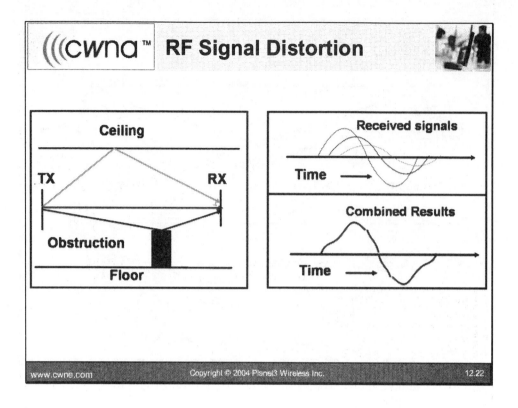

When reflected signals and the main signal are combined, distortion of the main signal can result. Distortion may mean that the transmitter has to resend.

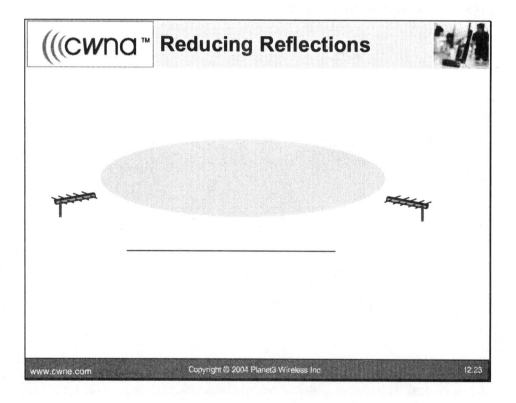

Use higher gain, less elevation beamwidth antennas or aim antennas upward to use the bottom of the pattern to connect. Using this technique will help prevent portions of the signal from bouncing off the ground reflector.

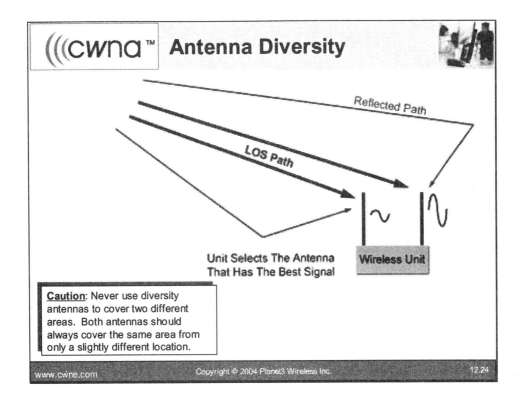

When reflected signals and the main signal are combined, the main signal's amplitude may be decreased or corrupted. Listening with multiple antennas increases the chance of receiving a high quality signal.

Chapter 13
Frequency Hopping Spread Spectrum (FHSS)

Objectives

Upon completion of this chapter you will be able to:

- Explain FHSS Technology
- Describe data rates and expected throughput with FHSS systems
- Explain FCC Rules regarding use of FHSS technology
- Explain FHSS Co-location techniques and advantages
- Explain how narrowband interference affects FHSS systems

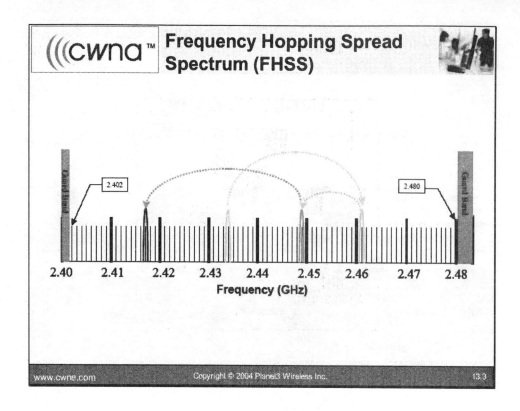

FHSS uses a large number of very "narrow" (low-bandwidth) channels. The signal hops between frequency channels in a pre-defined pattern. The pattern is known by both sender and receiver and is determined by a mathematical algorithm. The signal only stays in one place for a short time; therefore, it only interferes with any one frequency for a short time. To other transmitters on that frequency, the Spread Spectrum signal just looks like a tiny blip of noise. Some data may be lost, but only a small amount, since the "blip" is so short. The Spread Spectrum transmitter and receiver can recover the signal because they know what channel it will be on.

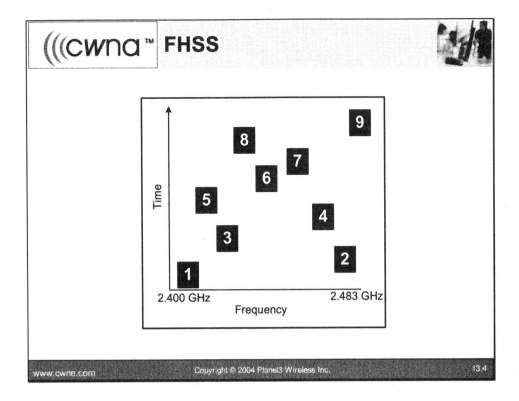

Number of FHSS Channels

Different FHSS algorithms define a different number of channels. More channels offer more options for avoiding interference, but also require more expensive, complex equipment. 802.11 uses 79 carriers of 1 MHz each.

Dwell Time

Dwell time is the amount of time that the transmitter stays on a single carrier before "hopping" to the next carrier in the algorithm. A shorter dwell time results in increased interference resistance, but results in less throughput. 802.11 has a maximum dwell time of 400ms (0.4 seconds) per 30 second period per carrier frequency.

Hopping Pattern

The hopping pattern is the pattern a station uses to hop through the carriers – this is called a "channel." The hopping pattern is Pseudo-random, but is designed to minimize interference. For example, 802.11 always hops at least 6 carriers at a time so that if there is interference on one carrier, the next hop will probably avoid that interference.

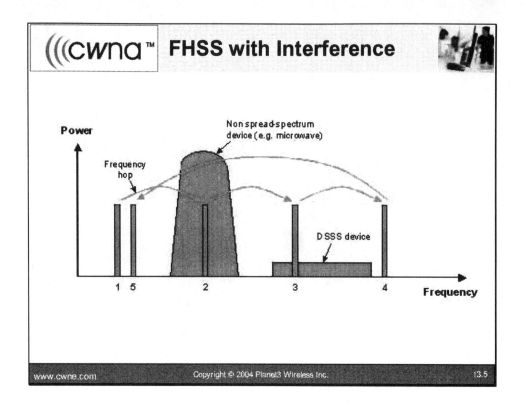

What if interference occurs on a given carrier and data is lost due to this interference? FHSS deals with data loss due to interference by automatically re-transmitting the lost data on the next hop. Since FHSS hops around so much and so quickly, acceptable data rates can be achieved even if some carriers experience interference. The 802.11 standard specifies FHSS at the 1 and 2 Mbps data rates. What if two pair of FHSS communicators choose the same channel at the same time? Interference will occur and data will be lost. The pseudo-random hopping algorithm minimizes the probability that the transmitters will stay on the same channel once they hop. For example, two devices may both hop to channel 56, but on the next hop, one device will go to one channel and the other will likely go to a different one. The 802.11 standard defines three groups of 26 unique hopping patterns, called "channels".

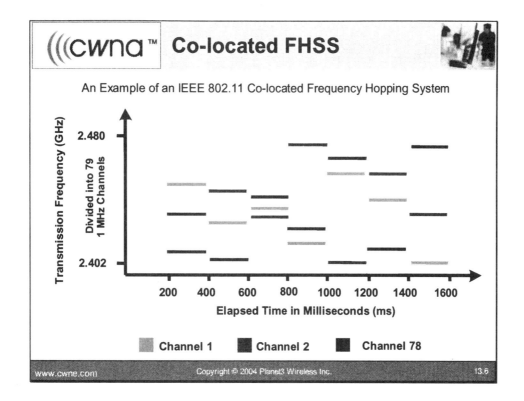

A co-located system is two or more transmitters whose coverage areas overlap; for example, two APs in the same area. Co-located systems are likely to occasionally hop onto the same carrier frequency at the same time causing data collisions. FHSS systems can be co-located using different channels for increased system throughput. Studies show that FHSS can support up to approximately 15 co-located systems with acceptable performance. Of course, this theoretical performance depends on many environmental variables.

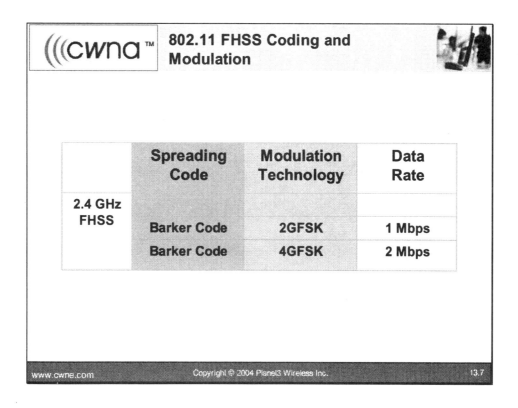

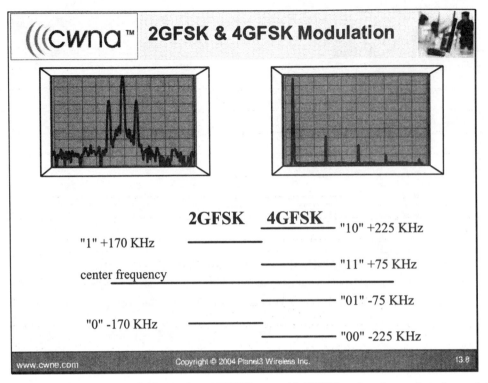

When using Gaussian Frequency Shift Keying (GFSK), as with FHSS technology, data is modulated (encoded using a carrier frequency) by shifting the frequency to predetermined points above and below a center frequency.

The FCC specified a minimum of 75 hops per channel prior to September, 2000. Under these rules, the Intentional Radiator was allowed to emit up to 1 Watt. In September of 2000, the FCC implemented a new rule allowing less than 75 hops per channel. Power output maximum was decreased to 125mW at the Intentional Radiator. Channel Bandwidth x Hop Count must equal at least 75 (i.e. 5 MHz x 15 Hops = 75). A 15 Hop minimum was put in place.

Chapter 14
Direct Sequence Spread Spectrum (DSSS)

Objectives

Upon completion of this chapter you will be able to:

- Explain DSSS Technology
- Describe data rates and expected throughput with DSSS systems
- Explain co-channel and adjacent channel interference with DSSS systems
- Explain DSSS Co-location techniques and advantages

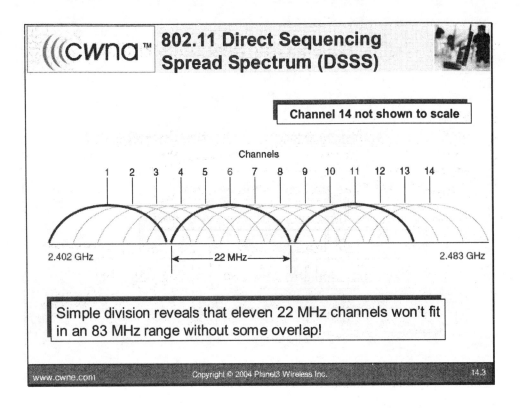

IEEE 802.11b defines 14 DSSS channels, 11 of which may be used in the U.S. Each channel is 22 MHz wide. The 2.4 GHz ISM Band is 83.5 MHz wide. The channel that we assign represents the center of the frequency range used by the DSSS transmitter. The signal will actually "bleed over" into neighboring channels. Bleed over is about 2 channels in either direction. If an access point is on channel 6, you will often also be able to capture its data on channels 4, 5, 7, and 8, but at decreased signal strength. Variables such as proximity, output power, and receiver sensitivity determine whether or not bleed over can be seen. Up to 3 non-overlapping channels can fit into the 2.4 GHz range.

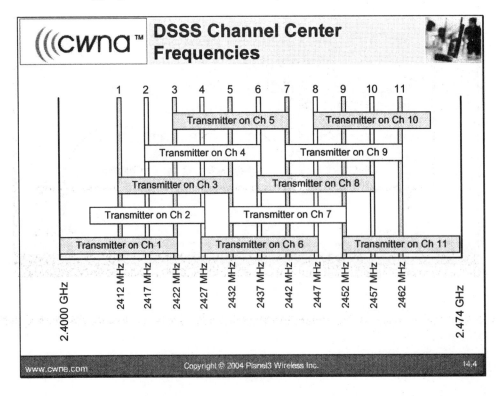

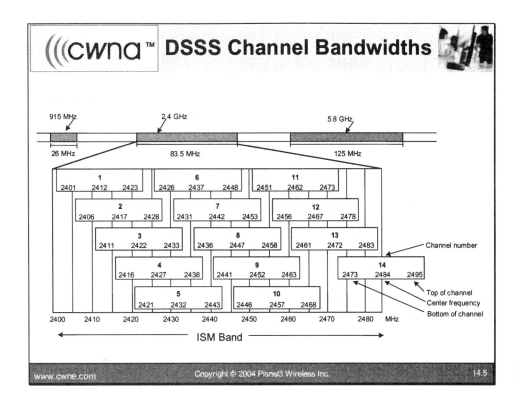

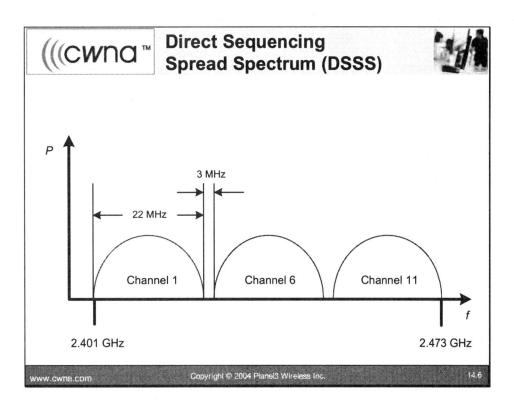

Each channel is 22 MHz wide. A maximum of three channels may be co-located (as shown) without overlap.

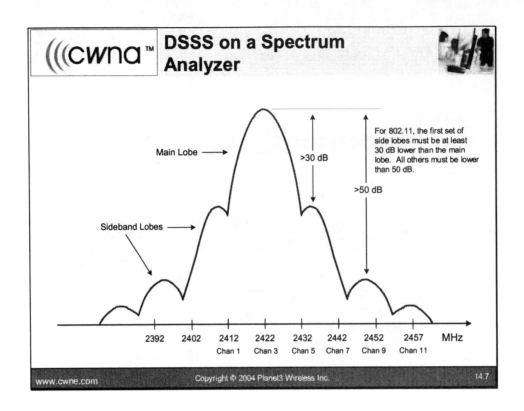

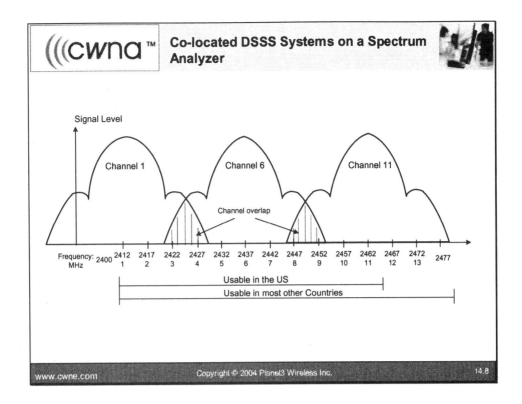

When using overlapping channels, sufficient distances between access points need to be observed to allow good operation for both access points. Non-overlapping channels still overlap somewhat due to sidebands. More channel separation is better (i.e., using channels 1 and 11 is better than using channels 1 and 6). Output power significantly increases adjacent channel interference.

How DSSS Deals with RF Interference

DSSS effectively transmits each bit multiple times, over multiple frequencies, which increases the probability that one of the transmitted bits will get through.

What if interference occurs and "wipes out" an entire DSSS channel?

In this case, none of the transmitted bits will get through. The DSSS transmitter will completely lose signal. DSSS is more susceptible to this type of interference than FHSS.

What if a pair of DSSS communicators use overlapping frequency ranges?

If the frequency ranges do not totally overlap, some of the bits may get through. If the frequency ranges overlap sufficiently, both communicators may partially or completely lose signal. It is best to always configure DSSS communicators using non-overlapping channels.

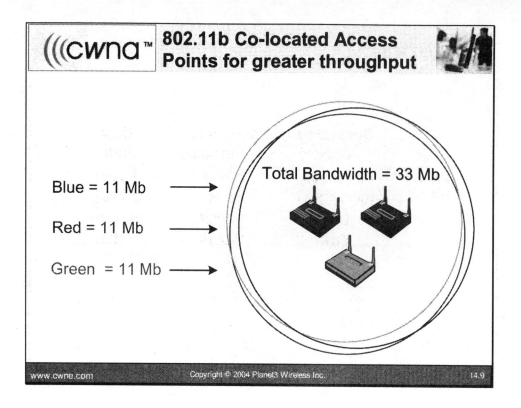

Using 802.11b, you can co-locate up to 3 systems with 11 Mbps bandwidth each.

Using 802.11g, you can co-locate up to 3 systems with 54 Mbps bandwidth each.

Using 802.11a, you can co-locate up to 8 systems with 54 Mbps bandwidth each.

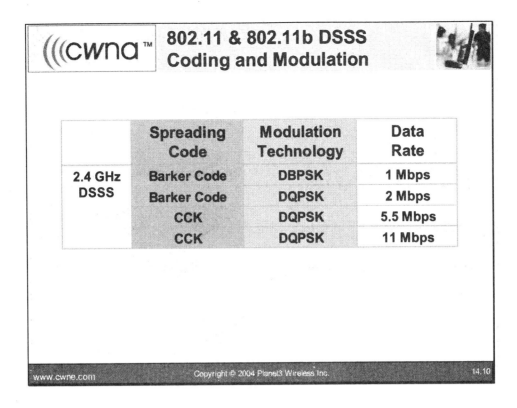

Data Rate (or bit rate) is expressed in Mbps, and relates to the MAC frame data only (not the PLCP header). The data rate is determined by modulation technology:

- DBPSK - 1 Mbps
- DQPSK - 2 Mbps
- CCK - 5.5/11 Mbps

MAC Management frames and multicast frames are transmitted at a lower data rate to be able to reach stations with different speed capabilities.

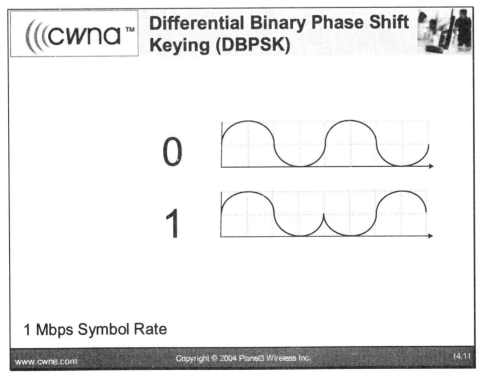

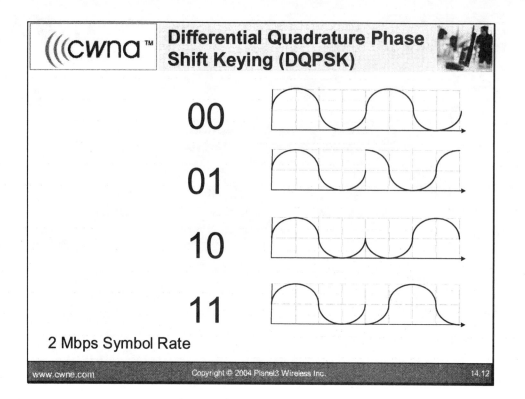

Chapter 15
Orthogonal Frequency Division Multiplexing (OFDM)

 **Objectives**

Upon completion of this chapter you will be able to:

- Understand OFDM and its use in 802.11a and 802.11g
- Understand the coding and modulation techniques used by the 802.11a and 802.11g standards

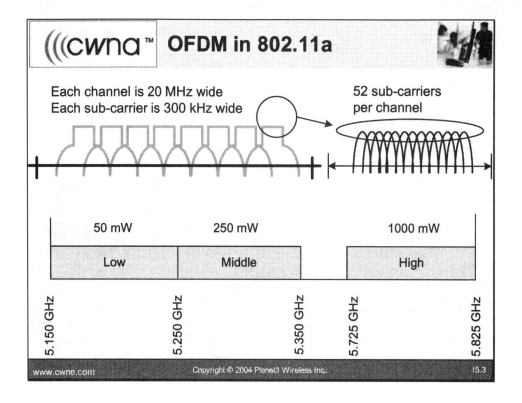

OFDM is used in 802.11a and 802.11g. OFDM allows more noise resistance and higher data rates in a given frequency range than FHSS and DSSS. The frequency range is divided into channels, similar to DSSS.

Each channel is divided into many small "sub-carriers". A single transmitter transmits and receives on all sub-carriers at once. Each sub-carrier individually has a low data rate, allowing for increased reliability, but in aggregate, high data rates are achieved. The frequencies of the sub-carriers overlap, but the signals are transmitted in such a way that the receiver can differentiate one sub-carrier from another.

802.11a uses three frequency bands in the 5 GHz range, which are known as the "low," "middle," and "high" bands or UNII-1, UNII-2, and UNII-3 respectively. Each band is 100 MHz wide. Even without OFDM, the 5 GHz ranges have greater potential for data transfer, since they have 300 MHz of bandwidth, whereas the 2.4 GHz range only has 83.5 MHz. Each frequency band has a different maximum power output. Low and middle are intended for indoor, medium-range use. All low-band devices are required to have an integrated antenna. High band is intended for long-range use (i.e. outdoor point-to-point links).

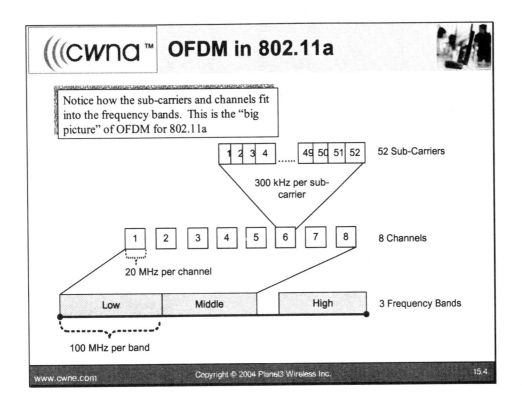

802.11a defines 8 non-overlapping channels in the lower two (lower and medium) bands. Each channel is 20 MHz wide. Each channel is subdivided into 52 sub-carrier channels, of approximately 300 kHz each. 48 of 52 are used for data, 4 of 52 are used for pilot tones. Each channel has a maximum data rate of 54 Mbps, but lower data rates may be used for increased range. A single station may transmit on each channel at a time. This process is similar to DSSS in which there are have three non-overlapping channels and one station at a time may transmit on any single channel. However, DSSS does not sub-divide the channels as OFDM does. When a station transmits on a channel, it sends data on all of that channel's sub-carriers simultaneously. The sub-carrier data is reconstructed at the receiver.

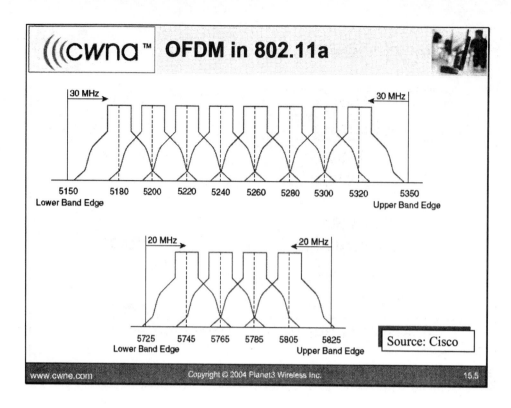

Unlike 802.11b DSSS and 802.11g OFDM, the channels don't overlap in 802.11a. It is the sub-carriers of each 802.11a channel that overlap. Adjacent channel interference can still occur with 802.11a at close range and high power. Since the channels do not overlap, all of them can be in use simultaneously. Therefore, 802.11a supports up to eight indoor, co-located systems.

Coding Technique	Modulation Technology	Data Rate
OFDM	DBPSK	*6 Mbps
OFDM	DBPSK	9 Mbps
OFDM	DQPSK	*12 Mbps
OFDM	DQPSK	18 Mbps
OFDM	16QAM	*24 Mbps
OFDM	16QAM	36 Mbps
OFDM	64QAM	48 Mbps
OFDM	64QAM	54 Mbps

* Mandatory support for these rates

802.11g Operational Modes

Mode	Rates (Mbps)
*ERP-DSSS	1,2
*ERP-CCK	5.5,11
*ERP-OFDM	6,9,12,18,24,36,48,54
ERP-PBCC	5.5,11,22,33
DSSS-OFDM	6,9,12,18,24,36,48,54

* Mandatory support for these modes

802.11g specifies several different signal transmission and encoding methods, each one offering different compatibility and performance characteristics. One of 802.11g's options is to use OFDM in the 2.4 GHz range. This option provides data rates of up to 54 Mbps. 802.11g radios also are required to support the 802.11b DSSS standard, for backwards compatibility with 802.11b radios. An 802.11b radio connecting to an 802.11g access point will not suddenly become capable of 54 Mbps data transfer; higher data rates are only possible when both radios support 802.11g.

Chapter 16
Packet Binary Convolutional Coding (PBCC)

Objectives

Upon completion of this chapter you will be able to:

- Explain PBCC's role in wireless LANs
- Explain the advantages and disadvantages of PBCC
- Explain how PBCC competes against DSSS and OFDM
- Explain PBCC's role in the 802.11g standard

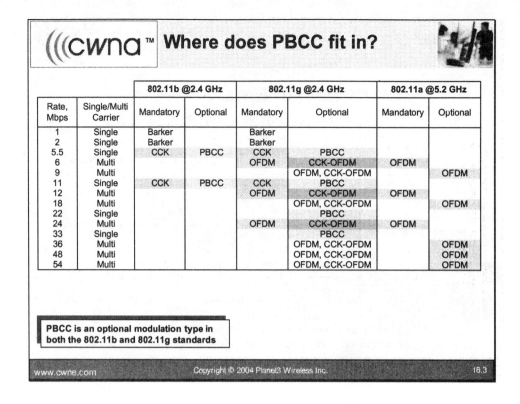

PBCC was developed as a method of increasing 802.11b connection speeds to 33 Mbps while maintaining backwards compatibility and peaceful co-existence with standard 802.11b devices.

PBCC is sometimes referred to as 802.11b+ in the market (using 22 Mbps). PBCC uses the same header as 802.11b so that the "duration" field can be decoded by legacy 802.11b devices. PBCC enables a maximum speed of 33 Mbps. Both 22 Mbps and 33 Mbps are supported in the 802.11g standard.

Advantages/Disadvantages of PBCC

- PBCC is faster than CCK
- PBCC may be faster than many speed settings of OFDM in a mixed mode environment because it does not require protection mechanisms to interoperate with CCK systems (802.11g requirement)
- Typical OFDMoperational speeds of 36 Mbps and 45 Mbps may have less throughput than PBCC-22
- PBCC allows for longer range at lower power than OFDM systems

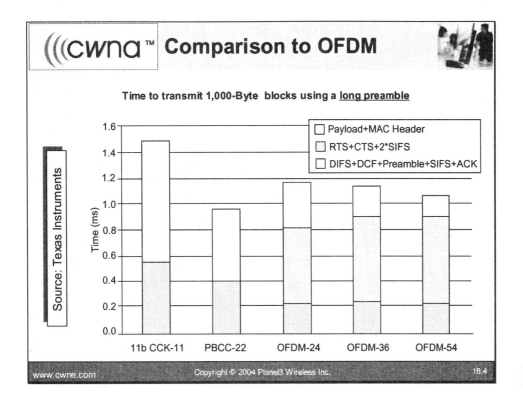

Texas Instruments (TI) argues that PBCC is typically faster than OFDM due to the overhead of OFDM protection mechanisms.

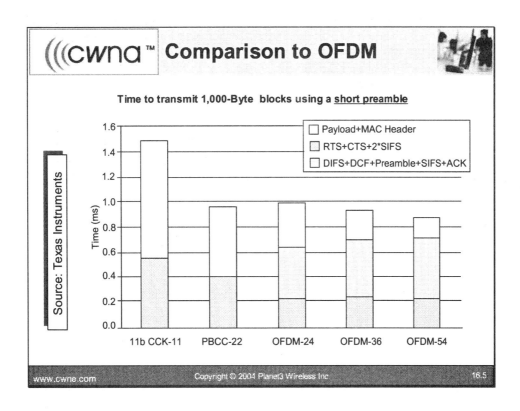

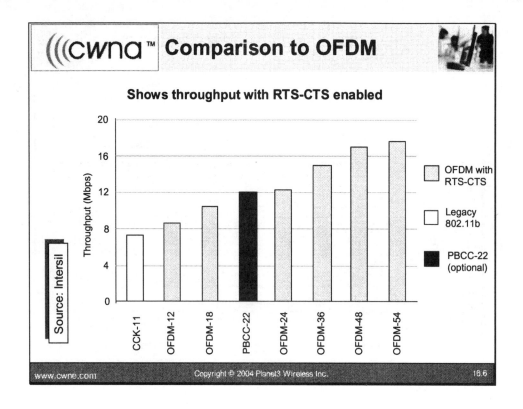

Intersil argues that even with the overhead of RTS/CTS for backwards compatibility, throughput is still higher in OFDM systems. The illustration shows throughput with RTS-CTS enabled.

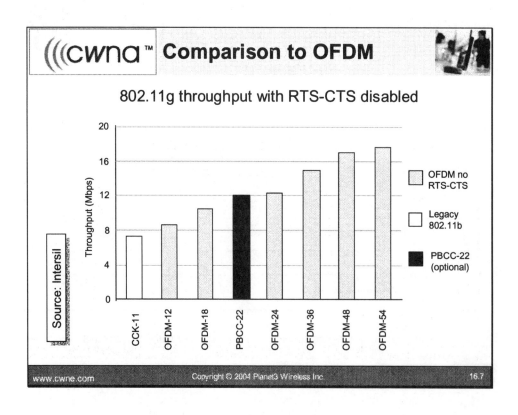

The illustration shows 802.11g throughput with RTS-CTS disabled.

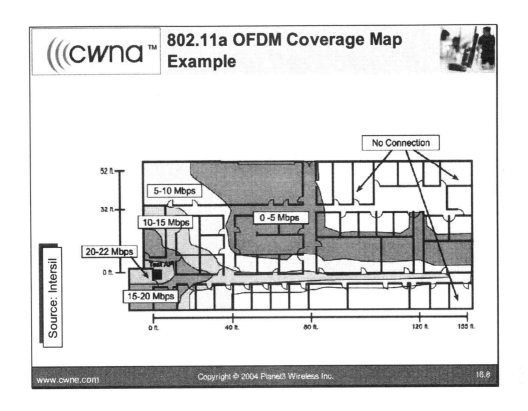
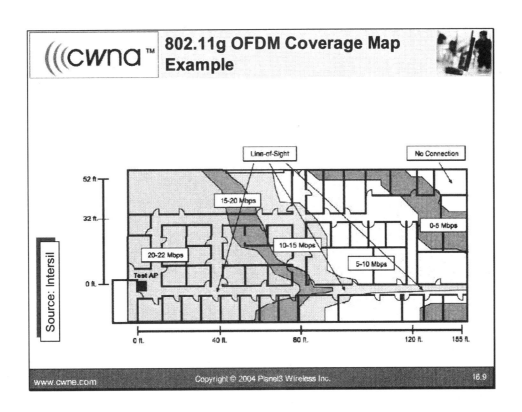

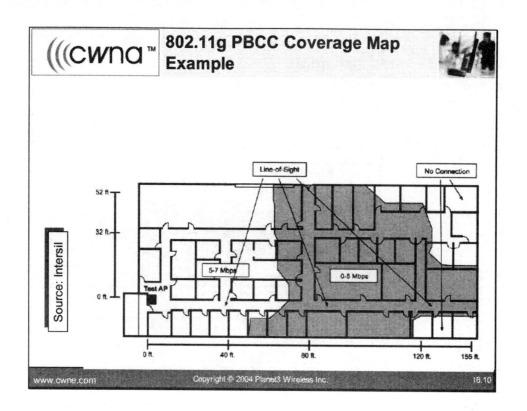

Chapter 17
Channels, Data Rates, Ranges, and Comparisons

Objectives

Upon completion of this chapter you will be able to:

- Describe and differentiate the general characteristics of the 802.11 series of connectivity standards
- Describe Dynamic Rate Shifting functionality
- Explain the differences between data rates and throughput on a wireless medium for each of the 802.11 series of connectivity standards
- Describe the channels for each of the license-free bands used with each 802.11 connectivity standard
- Compare 802.11 standards for:
  - Cost & Co-location considerations
  - Narrowband Interference Resilience
  - Data Rates and Throughput
  - Signal Range

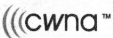

802.11 Connectivity Standards Summary

	802.11	802.11a	802.11b	802.11g
Frequency	2.4 GHz	5 GHz	2.4 GHz	2.4 GHz
Data Rates	1, 2 Mbps	5, 9, 12, 18, 24, 36, 48, 54 Mbps	1, 2, 5.5, 11 Mbps	6, 9, 12, 18, 24, 36, 48, 54 Mbps
S.S. Type	FHSS, DSSS	OFDM	DSSS	OFDM
Effective Data Throughput	1 Mbps	19 Mbps	5 Mbps	19 Mbps
Advertised Range	300 feet	225 feet	300 feet	300 feet
Encryption Available	Yes	Yes	Yes	Yes
Encryption Type	40-bit, 104-bit RC4	40-bit, 104-bit RC4	40-bit, 104-bit RC4	40-bit, 104-bit RC4
Provides Authentication?	No	No	No	No
Network Support	Ethernet (IEEE 802.3)	Ethernet (IEEE 802.3)	Ethernet (IEEE 802.3)	Ethernet (IEEE 802.3)

Radio Dynamics

- If you want <u>higher data rates</u>
  - You get shorter transmission range
- If you want <u>higher power output</u>
  - You get increased range
  - You get shorter battery life
- If you use <u>higher frequency radios</u>
  - You typically get more available frequency bandwidth
    - Increased bandwidth yields higher data rates
  - You get shorter range

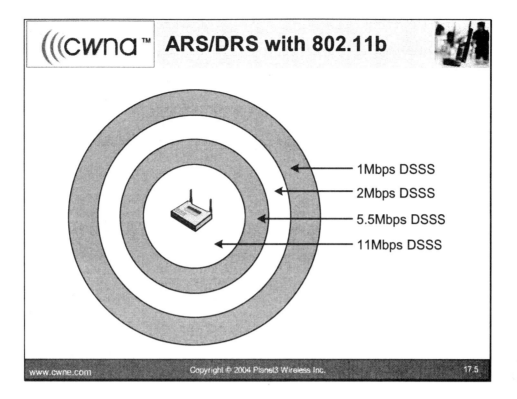

Adaptive (or Automatic) Rate Selection (ARS) & Dynamic Rate Shifting (DRS) are both terms used for providing a method of speed fallback on a wireless LAN client as distance increases from the access point.

ARS (or DRS) Process

1. Start at highest possible data rate
2. Fall-back to next lower data rate when 2 subsequent transmissions fail (ACKs missed)
3. Upgrade to next higher data-rate after 10 successful transmissions (ACKs) after 10 seconds

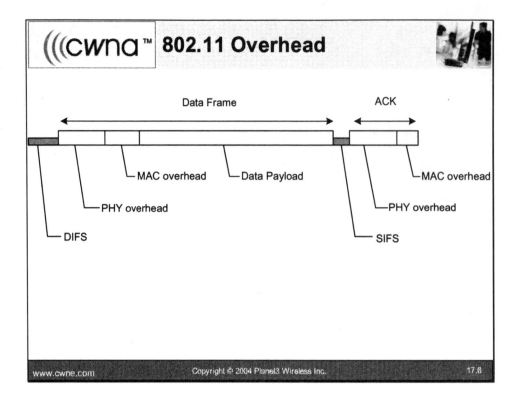

Throughput is always lower than data rate (bit rate) because of the following factors:

- Inter-frame spacing in the media
- PHY Overhead (Preamble)
- MAC Header overhead
- Need for of ACKs
- IEEE 802.11 Management & Control frames transmit at lower data rate
- Other users that share the media (half-duplex)
- Distance between sending station and access point or receiving station
- Environment (walls, reflecting objects)
- Type of operation (IBSS vs BSS/ESS)
- Contention window (CSMA/CA)
- Sources of interference
- Higher level protocols
- Encryption protocols

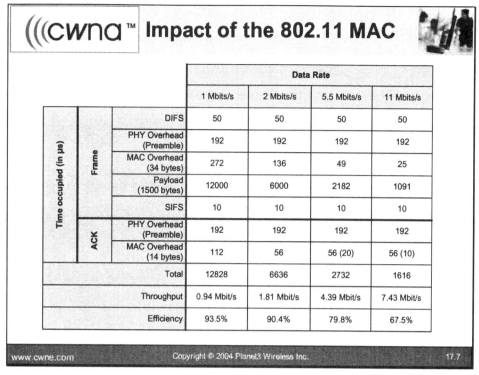

The total throughput numbers at the bottom of the chart on the preceding page represent the maximum MAC throughput, which is unrealistic, because it does not include interference, other users, etc.

Fixed length elements (independent of data-rate) become more significant at higher data-rates. For example, using DSSS technology:

- DIFS - 50 msec
- SIFS - 10 msec
- Preamble - 192 msec (based on a 144 byte "long" preamble sent at the minimum data rate)

MAC header for control and management frames can be shorter in small cells:

- at 5.5 Mbit/s reduced from 56 us to 20 msec
- at 11 Mbit/s reduced from 56 us to 10 msec

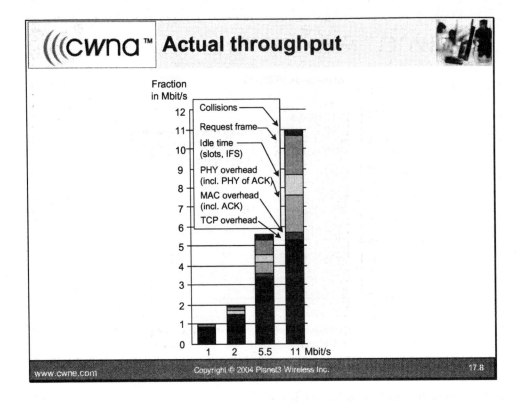

Realistic maximum numbers after inclusion of other factors:

- Protocol headers
- Management & Control frames
- Idle frames
- Interframe Spacing
- Collisions on medium
- Defer timers
- Performance of the actual computers

Throughput in a single BSS is lower than in an IBSS or ESS as result of the intra-cell relay function, in which traffic travels twice through the medium, invoking defers as part of CSMA/CA.

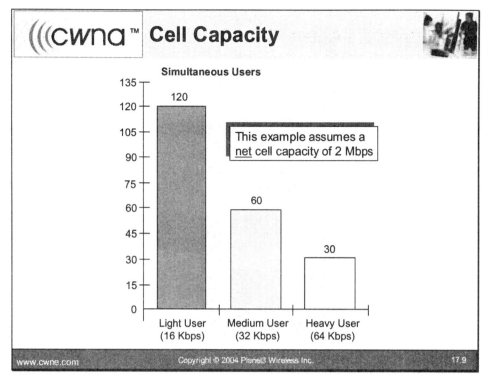

The number of stations per "radio cell" depends on:

- Bandwidth requirements per station, depends per application:
- Transaction processing: < 8 Kbit/sec
- Office Automation: < 64 Kbit/sec (depending on user profile)
- Multimedia: 100-800 Kbit/sec
- CAD/CAM: >1.5 Mbit/sec
- Total available bandwidth per cell
- Protocol
- Standards support
- Number of nodes
- Distance of each node

The number of co-located cells can be increased by using additional channels.

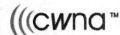

802.11b Channel Sets

Channel ID	FCC (U.S.) (GHz)	ETSI (Europe) (GHz)	MKK (Japan) (GHz)	IL (Israel) (GHz)
1	2.412	2.412	2.412	—
2	2.417	2.417	2.417	—
3	2.422	2.422	2.422	—
4	2.427	2.427	2.427	2.427
5	2.432	2.432	2.432	2.432
6	2.437	2.437	2.437	2.437
7	2.442	2.442	2.442	2.442
8	2.447	2.447	2.447	2.447
9	2.452	2.452	2.452	—
10	2.457	2.457 [1]	2.457	—
11	2.462	2.462 [1]	2.462	—
12	—	2.467 [1]	2.467	
13	—	2.472 [1]	2.472	—
14	—	—	2.484	—

Note 1: France is restricted to these four channels.

Source: Proxim

802.11g Channel Sets

Channel ID	FCC (U.S.) (GHz)	ETSI (Europe) (GHz)	MKK (Japan) (GHz)	IL (Israel) (GHz)
1	2.412	2.412	2.412	—
2	2.417	2.417	2.417	—
3	2.422	2.422	2.422	—
4	2.427	2.427	2.427	2.427
5	2.432	2.432	2.432	2.432
6	2.437	2.437	2.437	2.437
7	2.442	2.442	2.442	2.442
8	2.447	2.447	2.447	2.447
9	2.452	2.452	2.452	—
10	2.457	2.457 [1]	2.457	—
11	2.462	2.462 [1]	2.462	—
12	—	2.467 [1]	2.467	
13	—	2.472 [1]	2.472	—
14	—	—	2.484	—

Note 1: France is restricted to these four channels.

Source: Proxim

802.11a Channel Sets

Frequency Band	Channel ID	FCC (U.S.) (GHz)	ETSI (Europe) (GHz)	MKK (Japan) (GHz)	SG (Singapore) (GHz)
Lower Band (UNII-1)	34	—	—	5.170	—
	36	5.180	5.180	—	5.180
	38	—	—	5.190	—
	40	5.200	5.200	—	5.200
	42	—	—	5.210	—
	44	5.220	5.220	—	5.220
	46	—	—	5.230	—
	48	5.240	5.240	—	5.240
Middle Band (UNII-2)	52	5.260	5.260	—	—
	56	5.280	5.280	—	—
	60	5.300	5.300	—	—
	64	5.320	5.320	—	—

Source: Proxim

802.11a Channel Sets (cont'd)

Frequency Band	Channel ID	FCC (U.S.) (GHz)	ETSI (Europe) (GHz)	MKK (Japan) (GHz)	SG (Singapore) (GHz)
HIPERLAN Band	100	—	5.500	—	—
	104	—	5.520	—	—
	108	—	5.540	—	—
	112	—	5.560	—	—
	116	—	5.580	—	—
	120	—	5.600	—	—
	124	—	5.620	—	—
	128	—	5.640	—	—
	132	—	5.660	—	—
	136	—	5.680	—	—
	140	—	5.700	—	—
Upper Band (UNII-3)	149	5.745	—	—	5.745
	153	5.765	—	—	5.675
	157	5.785	—	—	5.785
	161	5.805	—	—	5.805
ISM Band	165	5.825			5.825

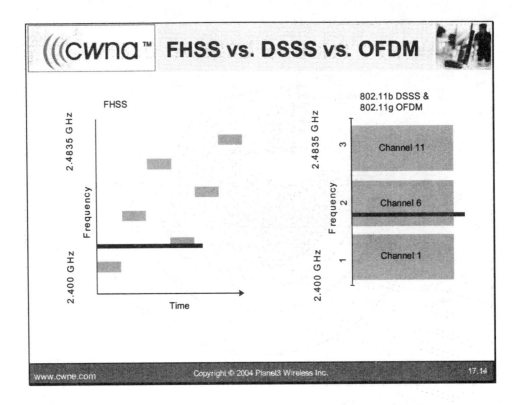

Narrowband Interference Resilience

- 802.11 FHSS = best
- 802.11a/g OFDM = better
- 802.11b DSSS = good

Costs

802.11g units are now replacing 802.11b units because of backwards compatibility and low cost. 802.11a has less range than 802.11b/g which requires more access points to take advantage of the higher speeds. A separate 5 GHz site survey is required in addition to the original 802.11b site survey. Some manufacturers support new security standards only in 802.11a & 802.11g cards, which may require firmware and software upgrades. FHSS systems are legacy, slow, expensive, and difficult to purchase, because few manufacturers still stock the units.

Data Rates

The data rate is dependent on the technology in use. FHSS systems have lower throughput. After a frequency "hop," the transmitter and receiver must take some time to synchronize. Data cannot be transmitted during this time. FHSS is specified at the 1 and 2 Mbps data rates.

DSSS systems have higher throughput because data is always being transmitted — there is no "hopping penalty". DSSS is specified at the 1, 2, 5.5, and 11 Mbps data rates. Interference in the 2.4 GHz range lowers effective throughput.

802.11g OFDM has higher throughput. If 802.11g cards are used exclusively, up to 54 Mbps can be achieved. The data rate will only be up to 11 Mbps when an 802.11g radio talks to an 802.11b radio. 802.11g devices are also susceptible to 2.4 GHz interference.

802.11a OFDM has the highest throughput of all, with data rates up to 54 Mbps, and is immune to 2.4 GHz interference because the 5 GHz range is much less crowded.

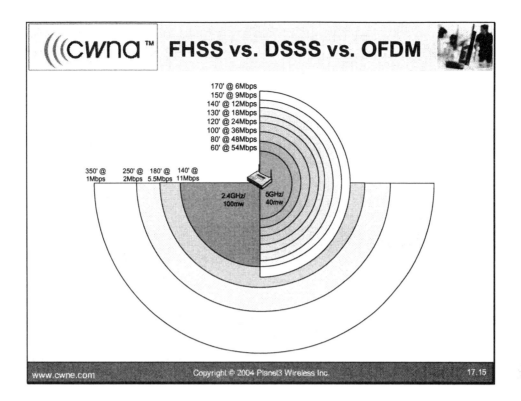

Range

High speeds of 802.11a systems are only good for a short distance indoors due to power regulations. 802.11a systems do not penetrate walls and other solid objects very well. Outdoor 802.11a systems are becoming popular, especially long-distance point-to-point systems.

802.11b/g systems have good range, but are subject to a more congested frequency band. 802.11b/g systems are more popular indoors because the 2.4 GHz frequencies penetrate walls well.

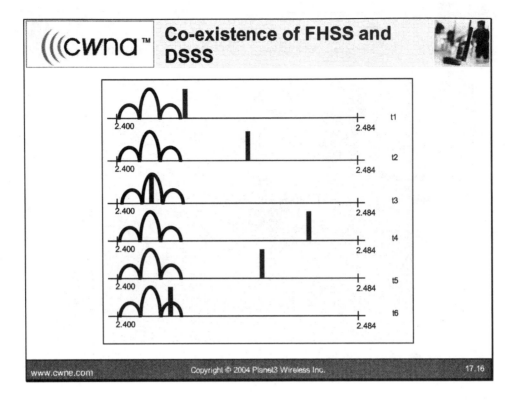

Often, FHSS and DSSS systems can co-exist without significant problems, but there will be some degradation in data rate and throughput. The amount of degradation depends on output power and proximity of the systems.

Co-location

802.11 FHSS systems can have a high number of co-located systems, but costs are high and throughput is low. 802.11b/g DSSS/OFDM systems only support 3 non-overlapping channels in the 2.4 GHz band. Distance between systems, antenna type, and output power all affect effective throughput when co-locating. 802.11a systems support 4 channels per band, with 2 consecutive bands for indoor use yielding 8 channels for use in co-location.

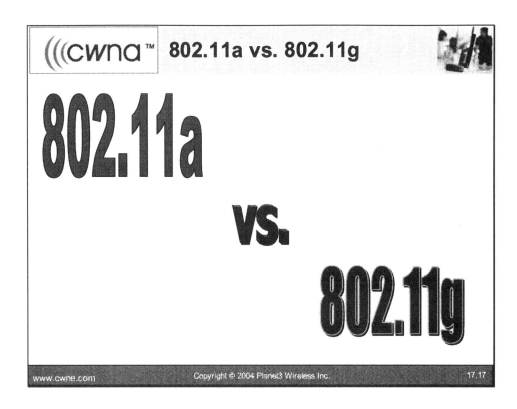

Factors to consider when comparing 802.11a to 802.11g

802.11a and 802.11g are specified at comparable data rates; however, it is expected that 802.11a will have higher effective throughput due to its operation in the less-crowded 5 GHz range.

802.11a frequencies will not travel as far, and will not penetrate obstacles as well as 802.11g frequencies.

802.11g will be backwards compatible with existing 802.11b networks (although not at the higher data rates).

802.11g may be preferable to 802.11a if:

- You already have a significant investment in 802.11b equipment and want the new cards to work with the old equipment.
- Your users occasionally need high throughput, but not always.
- You can install a single 802.11g access point as a "high-speed hotspot".
- Users can still connect to the existing 802.11b access points at lower speeds.

Chapter 18
Channel Reuse and Mixed Mode

Objectives

Upon completion of this chapter you will be able to:

- Explain the concept of channel reuse and its importance when using DSSS and OFDM systems
- Describe how cell dimensions relate to data rates
- Explain how load-balancing and hot-standby features work and their importance
- Describe 802.11b/g Mixed-mode problems

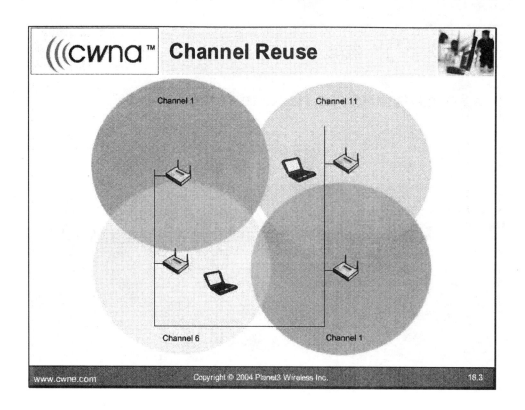
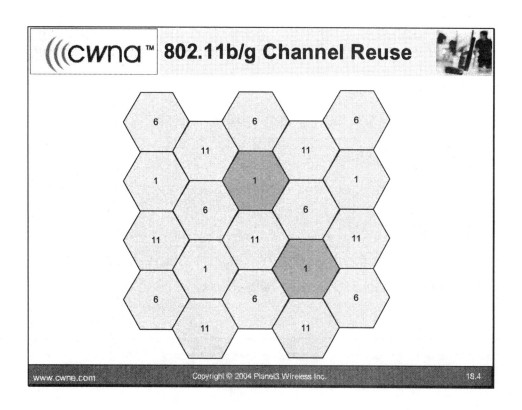

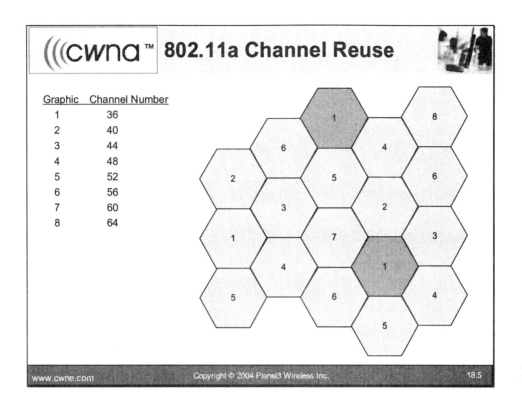
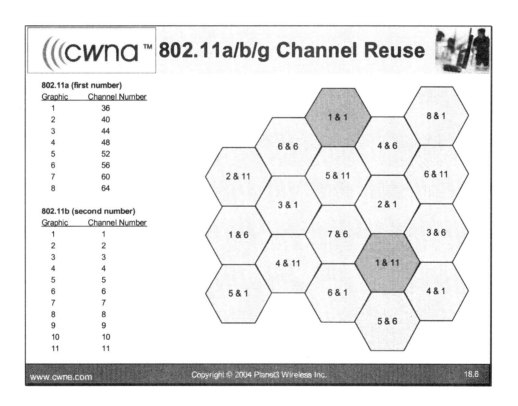

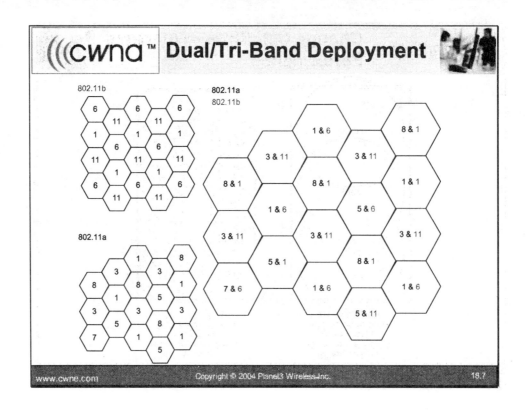
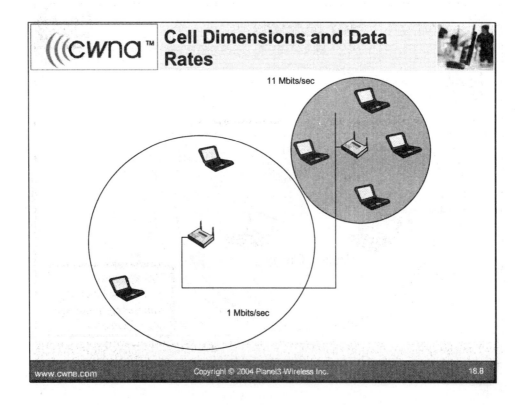

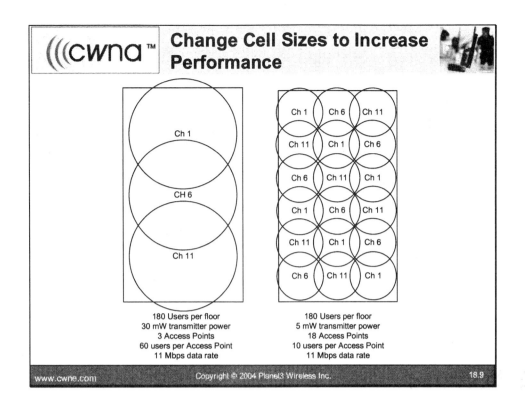

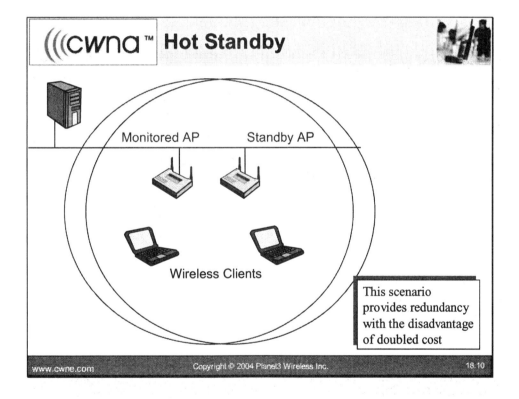

802.11b/g Mixed Mode (Downstream)

#802.11b Clients					Number of 802.11g Clients					
10	5.9	6.2	6.5	6.8	7.0	7.2	7.4	7.6	7.8	8.0	8.2
9	5.9	**6.2**	6.5	6.8	7.1	7.4	7.6	7.8	8.0	8.2	8.3
8	5.9	6.3	**6.6**	6.9	7.2	7.5	7.7	8.0	8.2	8.4	8.5
7	5.9	6.3	6.7	**7.1**	7.4	7.7	7.9	8.2	8.4	8.6	8.8
6	5.9	6.4	6.8	7.2	**7.6**	7.9	8.2	8.4	8.7	8.9	9.1
5	5.9	6.5	7.0	7.4	7.8	**8.2**	8.5	8.7	9.0	9.2	9.4
4	5.9	6.6	7.2	7.7	8.2	8.5	**8.9**	9.2	9.4	9.6	9.8
3	5.9	6.8	7.6	8.2	8.7	9.1	9.4	**9.7**	9.9	10.2	10.4
2	5.9	7.2	8.2	8.9	9.4	9.8	10.2	10.4	**10.7**	10.9	11.1
1	5.9	8.2	9.4	10.2	10.7	11.1	11.3	11.6	11.7	**11.9**	12.0
0	0.0	22.1	22.1	22.1	22.1	22.1	22.1	22.1	22.1	22.1	**22.1**
	0	1	2	3	4	5	6	7	8	9	10

Chart includes TCP/IP overhead, and denotes downstream packet flow only. Source: Intersil

The situation of CCK and OFDM radios operating on the same channel is analogous to the "hidden node" problem because the CCK radios cannot "hear" the OFDM transmissions. Through the use of a protection mechanism, OFDM radios are able to operate on the same channel as existing DSSS radios without collisions. The 802.11g standard specifies two such protection mechanisms:

- RTS/CTS
- CTS-to-Self

802.11g systems will only be as fast as 802.11a systems when there are no 802.11b devices present. In mixed mode (802.11b/g) operation, 802.11g OFDM packets are preceded by an RTS/CTS or a CTS-to-Self frame using 802.11b modulation (BPSK, QPSK, or CCK – based on the lowest configured data rate). The CTS conveys the length of time required for the ensuing high speed OFDM frame and ACK. 802.11b clients receiving this CTS will remain idle for the specified period of time, avoiding the collision with the OFDM packet.

802.11b/g mixed mode (downstream)

In a mixed 802.11b/g environment the number of 802.11g clients vs. 802.11b clients matter for throughput. More 802.11b clients mean less throughput overall. Downstream, the network bandwidth is shared. Every client gets the same number of packets regardless of access rate or modulation type because the access point alternates sending queued packets between nodes (round-robin). It takes *much* longer to send a 1500 byte packet to an 802.11b station than an 802.11g station.

802.11b/g Mixed Mode (Upstream)

- Mixed mode 802.11b/g results in approximately twice as many transmit opportunities for 802.11g clients due to statistically lower backoff counter values
  - This helps balance upstream/downstream airtime between 802.11b and 802.11g clients, but not much

> An informative whitepaper from AirMagnet on 802.11b/g mixed mode analysis can be found in the appendix of this course guide.

Access Point Modes

- Most 802.11g access points support:
  - *802.11b-only* mode using DSSS/CCK
  - *802.11g-only* mode using ERP-OFDM or DSSS-OFDM
  - *802.11b/g-mixed mode* using a combination of mandatory and optional modulations

Chapter 19
Arbitration, RTS/CTS, Carrier Sense, and Hidden Nodes

Objectives

Upon completion of this chapter you will be able to:

- Define and explain arbitration and its use in an RF medium
- Describe the differences between CSMA/CD and CSMA/CA
- Describe the modes of operation wireless LANs use to communicate
- Explain Carrier Sense Mechanisms used by 802.11
- Identify and resolve the issues with hidden nodes and Near/Far
- Explain how the Request-to-Send / Clear-to-Send (RTS/CTS) transmission protocol is used in wireless LANs

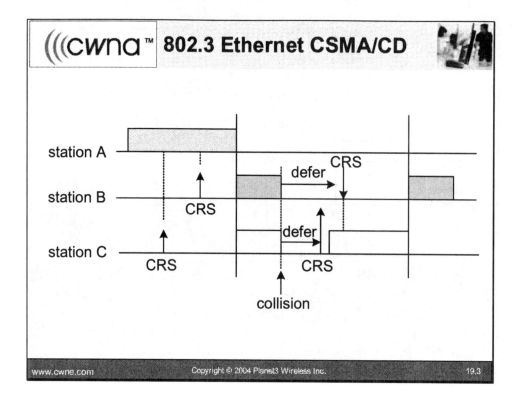

In every network, there must be some rule for determining when a station may access the medium and when it may not. These methods are referred to as access methods or arbitration methods. The 802.11 standard refers to them as coordination functions. Some access methods are contention-based, in which stations compete for network bandwidth. In contention-based systems, all stations may not get equal access to network bandwidth. Examples of contention-based access methods include CSMA/CD and CSMA/CA.

Some access methods are not contention-based, and every station gets equal access to network bandwidth. These networks often involve a central "management" device. Examples include token-passing and polling.

Adapters that can detect collisions (i.e. Ethernet adapters) use the following processes:

- Carrier sensing: listen to the media to determine if it is free
- Initiate transmission as soon as carrier drops
- When collision is detected, station defers
- When defer timer expires: repeat carrier sensing and start transmission

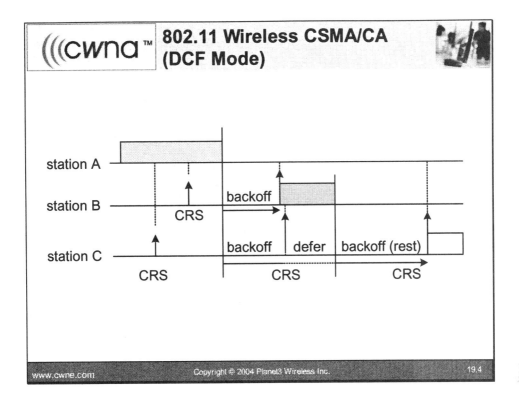

The fundamental access method of 802.11 is CSMA/CA, "carrier sense multiple access with collision avoidance." Wireless nodes cannot detect collisions so different coordination schemes must be used.

Wireless access methods are contention-based, using random backoff timers. 802.11 refers to this as a Distributed Coordination Function (DCF), referring to the coordination of which station can transmit and when. Coordination is distributed among all members of the BSS, and is designed to reduce the probability that a collision will occur just after transmission, when collisions are most likely.

 **CSMA/CA Behavior Summary**

- Carrier Sense

- Multiple Access

- Collision Avoidance

Carrier Sense means that the station determines if the medium is currently controlled by another station. If the medium is currently controlled by another station, then the station will defer to that station that is controlling the medium, and will not transmit. If not, then the station will take control of the medium, and proceed with transmission.

To ensure minimum interframe spacing (IFS), the network must be idle for a certain amount of time between frames.

In "Multiple Access", multiple stations may access the medium using collision avoidance. Instead of transmitting immediately after the network becomes available, each station will wait (back off) for a random amount of time. CSMA/CD only backs off when there has been a collision. CSMA/CA backs off whenever a station defers to another station.

Why not CSMA/CD?

- **C**arrier **S**ense **M**ultiple **A**ccess with **C**ollision **D**etection

- Transmitting stations cannot detect collisions

Using Carrier Sense Multiple Access with Collision Detection (CSMA/CD), stations sense the carrier and wait for an idle medium (as in CSMA/CA). While transmitting, the stations listen to the wire and detect the invalid signal caused by a collision. Wired specifications specifically guarantee that every station on the wire will be able to detect this invalid signal. In a wireless medium, there is no guarantee that the transmitting station will be able to detect the invalid signal caused by a collision, because radios are a half duplex medium. Stations may not be able to perceive each others' signals due to environmental obstructions. The station will probably not be able to detect the invalid signal due to the strength of its own good signal. As an illustration, if you are standing in a room shouting loudly, you probably will not be able to tell that someone else, 100 feet away but in the same room, is also talking in a normal tone.

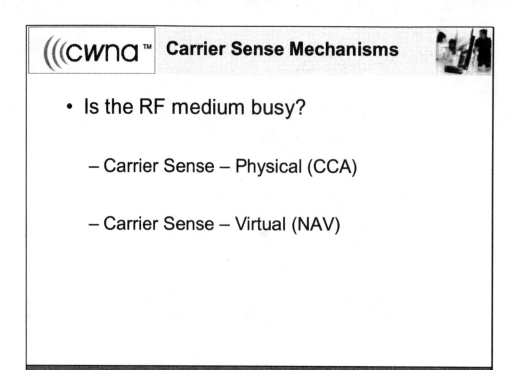

Methods by which a station can determine whether any other station is currently using the medium:

- Physical layer provides a carrier sense mechanism
- Involves detection of RF energy near the receiver
- Specifics vary
- May require detection of any energy at all
- May specifically require detection of 802.11 encoded signal
- Called "Clear Channel Assessment (CCA)"
- 802.11 MAC layer provides a "virtual" carrier sense mechanism
- Frames contain a Network Allocation Vector (NAV)
- Indicates how long a station will hold the medium
- Other stations defer until the NAV expires

Physical carrier sense involves a procedure called Clear Channel Assessment (CCA). CCA equates to the station asking the Physical Layer, "Is anybody else transmitting right now?" The Physical Layer performs CCA by measuring the RF energy at its receiver and using some pre-programmed logic to determine whether another station is currently transmitting. DSSS, FHSS, and OFDM physical layers each have different methods of determining the answer to this question. The specifics of these methods are less important to protocol analysts and administrators than they are to computer scientists and electrical engineers who design NICs.

Virtual Carrier Sense involves two methods of reserving the medium: RTS (Request to Send) and CTS (Clear to Send) frames. Prior to transmission of data frames, stations reserve the medium for a certain amount of time, which is the time it will take for their data to be transmitted and acknowledged. Stations set the Duration/ID field in the MAC header to indicate the length of time for which they are reserving the medium. Virtual carrier sense addresses the hidden node problem.

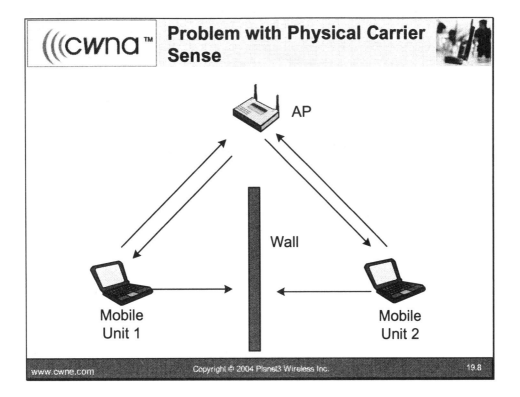

In a BSS, all stations can communicate with the access point, but all stations may not be able to see each other's signals. Stations 1 and 2 can both communicate with the access point, and they can communicate with each other through the access point. However, they cannot see each other's signals to perform physical carrier sense (CS). One station might be transmitting and the other would not hear the transmission. This situation is known as the hidden-node problem.

As an administrator, you should be aware of the implications of the hidden node problem on your activities. Because of this problem, you may not be able to see all of the traffic that is transmitted on the BSS with a wireless packet analyzer. It depends on the protocol analyzer's location relative to the access point and other transmitters. You may see responses with no requests or requests with no responses. In the wired LAN, missing frames would be extremely suspicious. In the wireless world, missing frames are not inherently a problem.

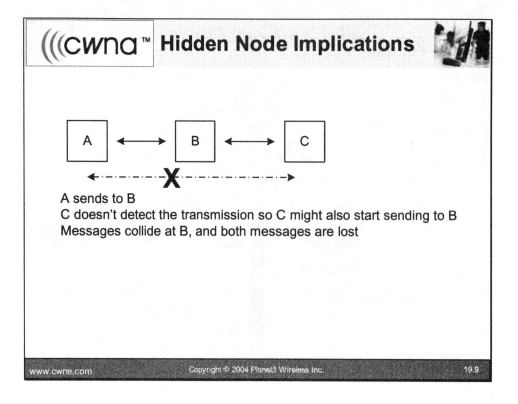

The best way to work around the hidden-node problem is to place a protocol analyzer as close as possible to either the access point or the client being investigated. This will ensure that the analyzer sees all of the traffic sent and received by these stations.

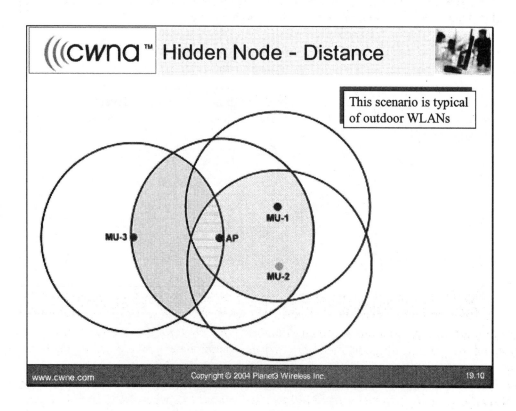

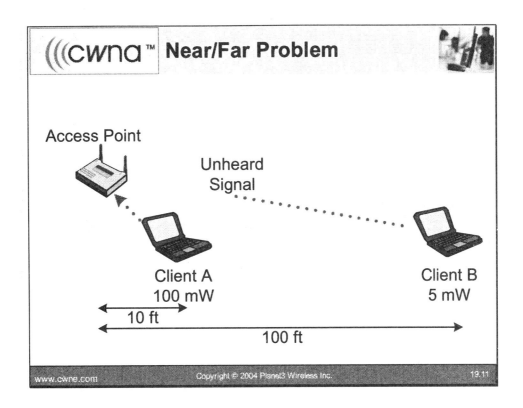

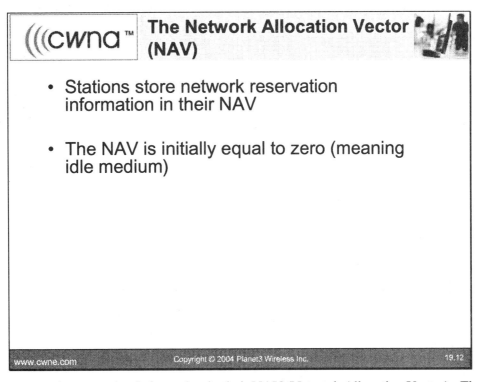

Stations store network reservation information in their NAV (Network Allocation Vector). The NAV is initially equal to zero (meaning idle medium). When a station reserves the medium, the NAV is set to the length of time for which the medium is reserved. The length of time is equal to however long it will take to transmit the entire data frame and receive its acknowledgement. In the case of a fragmented frame, the length of time is equal to however long it will take to transmit and acknowledge one fragment. Each subsequent fragment and acknowledgement acts as an implicit RTS/CTS exchange, resetting the NAV in non-transmitting stations. As long as the NAV is non-zero, the medium is considered to be busy. The NAV counts down at a constant rate until it is zero again. When/if the NAV equals zero, the network is considered idle again.

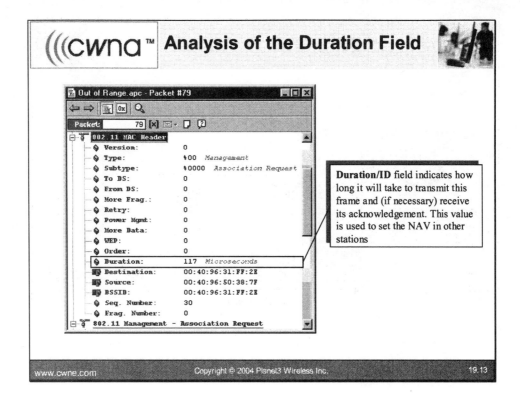

Stations reserve the medium by setting the Duration/ID field in either a RTS frame or a data frame. An RTS frame is used if the station has chosen to use RTS/CTS. A data frame is used if the station is not using RTS/CTS. A station updates its NAV with the value of the Duration/ID field when it receives a valid frame and:

- The new NAV value is greater than their current NAV value
- The frame is not addressed to the station

If stations updated their NAV after receiving a frame, they could not subsequently acknowledge the frame, since the NAV would indicate a busy medium. Each frame that is successfully sent (whether it uses RTS/CTS or not) reserves the medium for the appropriate length of time to transmit that frame and receive its acknowledgement. No other station will try to transmit during this time.

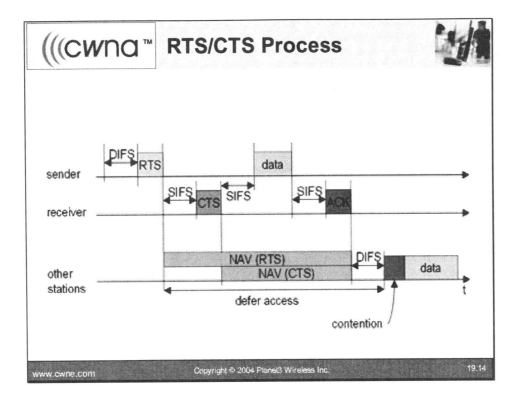

A station may choose to reserve the medium prior to transmission by exchanging RTS and CTS frames with the intended recipient of its data. The transmitter sends an RTS frame to the recipient. The recipient answers with a CTS. The RTS/CTS method imposes significant overhead, especially when small packets are being sent, and may not be appropriate for every situation.

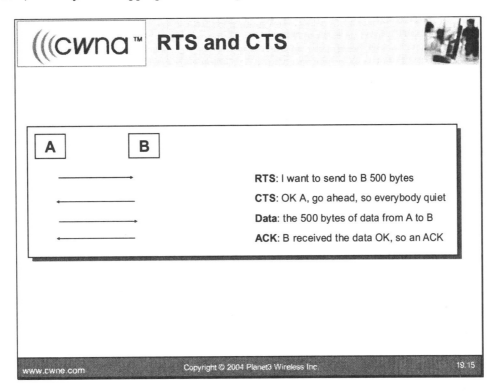

A threshold is configured in each station that determines whether RTS/CTS will be used for each packet. If a packet is longer than the threshold, the station will use RTS/CTS. If not, the station will transmit the data using normal CSMA/CA. Administrators may set this threshold to 0 to force RTS/CTS for all frames. Administrators may set this to a large value to disable RTS/CTS.

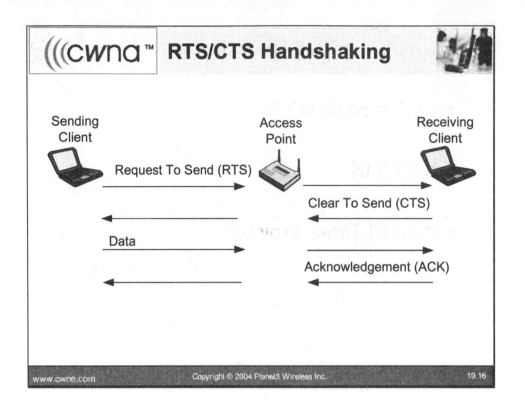

Carrier Sense Summary

- CCA = positive?

- NAV = 0?

- Backoff Timer expired?

Before transmitting a frame, a station must Sense Carrier and confirm that the network is idle. This process consists of two parts:

1. Check the NAV - If the station's NAV is greater than zero, the network is considered to be busy
2. Perform CCA - Physical layer reports on the idle/non-idle status of the medium by measuring RF energy near its antenna

Specific criteria for idle/non-idle vary between different physical layers. If both CCA and NAV report an idle network, then the station is allowed to transmit provided its backoff timer is equal to 0. If network is not idle, then the station defers to the station that is currently controlling the medium. When the medium becomes idle, the station waits the appropriate Interframe Space (IFS) before transmission. If the medium is idle and has been idle for DIFS, then the station transmits. The station then waits for acknowledgement. If no acknowledgement, the station queues the frame for retransmission. When deferring, the backoff timer is chosen after IFS. While medium is idle, the backoff timer is counting down. If medium becomes busy, the backoff timer pauses. When the backoff timer reaches zero, the station transmits.

The backoff timer is guaranteed to reach zero at some point provided the medium is perceived to be idle at some point (which always happens).

If, after all that, a collision still occurs, it is okay, because the 802.11 MAC is reliable, and will retransmit.

Chapter 20
Beacons, Probes, and Scanning

Objectives

Upon completion of this chapter you will be able to:

- Explain what beacons are and what information they provide
- Explain probe request and response frames
- Explain passive and active scanning

Finding a BSS - Passive

- **Passive Scanning** method: Listen for Beacon Management frames (beacons)

Channel	Signal	Data Rate	Size	Relative Time	Protocol
1	100%	1.0	105	00.000000	802.11 Beacon
1	100%	1.0	105	00.000155	802.11 Beacon
1	94%	1.0	105	00.000308	802.11 Beacon
1	94%	1.0	105	00.000458	802.11 Beacon
1	100%	1.0	105	00.102215	802.11 Beacon
1	100%	1.0	105	00.204642	802.11 Beacon
1	100%	1.0	105	00.306871	802.11 Beacon
1	94%	1.0	105	00.409428	802.11 Beacon
1	100%	1.0	105	00.511949	802.11 Beacon
1	97%	1.0	105	00.614062	802.11 Beacon
1	94%	1.0	105	00.716690	802.11 Beacon

Before a station can join a BSS, the station must learn that a BSS exists by scanning. The passive scanning method is when the station is listening for Beacon Management frames (beacons). Access points periodically send beacon frames which contain the access point's SSID and other information. If the station hears a beacon frame with an SSID matching its configured SSID, it may issue an Association Request to the access point sending the beacon. The station may hear beacons from access points with a different SSID than the station is configured for, but it will not associate with these access points unless the station is configured to use any SSID.

Beacons

- Used as a type of "homing signal" so that stations can find access points
- Used for time synchronization
- Used to announce the presence of an access point
- Passes channel selection and network configuration information
- Used for FHSS, DSSS, & OFDM spread spectrum networks
- Cannot be turned off, but fields inside the beacon can be modified manually
- Sent at lowest common data rate
- Typically sent 10/second by default (configurable)

Finding a BSS - Active

- **Active Scanning** method: Send Probe Request frames

Channel	Signal	Data Rate	Protocol
1	14%	1.0	802.11 Probe Req
1	100%	1.0	802.11 Probe Rsp
1	100%	1.0	802.11 Probe Rsp
1	100%	1.0	802.11 Probe Rsp
1	17%	1.0	802.11 Probe Req
1	100%	1.0	802.11 Probe Rsp
1	100%	1.0	802.11 Probe Rsp

A station may discover a BSS by using the active scanning method, in which the station sends probe request frames, which contain the SSID that has been configured in the station. Any access point that hears the Probe Request and that has the same SSID as the SSID in the Probe Request sends a probe response back. The access point must be operating on the same channel that the probe was sent in order to hear the probe. The station may set the SSID to a special value (known as the broadcast SSID) to indicate that all access points should respond. The "broadcast SSID" is a blank or null value in the SSID field.

When a station is not associated with any access point, it may scan through all channels, looking for an access point on each of the channels. When the station is associated with an access point, it only searches on the channel that the access point is using.

If stations are receiving beacons or probe responses, the next logical step is to attempt to associate with one of the access points sending the beacons or probe responses. The SSID in the station must match the SSID in the access point. A station may be configured with a "null" SSID, meaning that it will associate with any access point. Some access points will be configured to reject stations with the null SSID. Other vendor-specific factors may dictate whether the station will attempt to associate or not. Stations may periodically send probe requests even after they have associated. These periodic probes may be used to find new access points or may be used to confirm that the current access point is still the best (strongest signal). Beacons/probe response frames may also be used to find "rogue" access points.

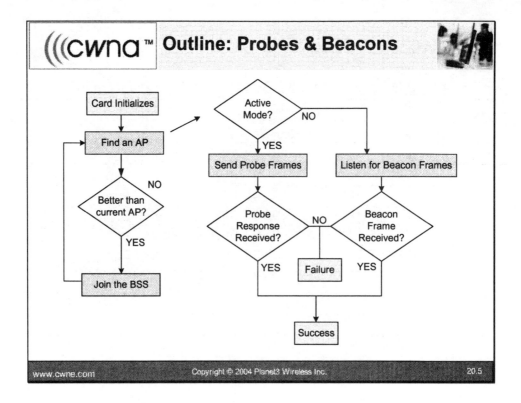

Commonly, a station will be in range of more than one valid access point and will have to choose one with which to associate. This can happen:

- When a station first starts up
- When a station is roaming out of one access point's coverage area into the overlapping coverage area of two or more access points
- When a station is sitting still, but, due to environmental factors, one access point's signal becomes weaker and another access point's signal becomes stronger.

The manner in which a station chooses between access points is not specified by the 802.11 standard. Individual vendors use different criteria to decide which access point is "best". Details are proprietary and generally kept secret.

Given two valid access points, it is hard to predict which one a card will prefer. Different vendor's cards may choose differently.

Chapter 21
Authentication, Association, and Roaming

Objectives

Upon completion of this chapter you will be able to:

- Explain the following processes and frames:
  - Authentication
  - Deauthentication
  - Association
  - Reassociation
  - Disassociation

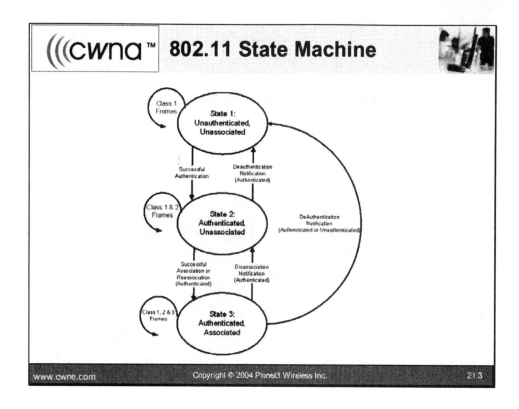

The 802.11 standard specifies something called a state machine. This diagram shows the various "states" that an 802.11 station can be in when moving from a state of "not connected" to a state of "connected" to an access point.

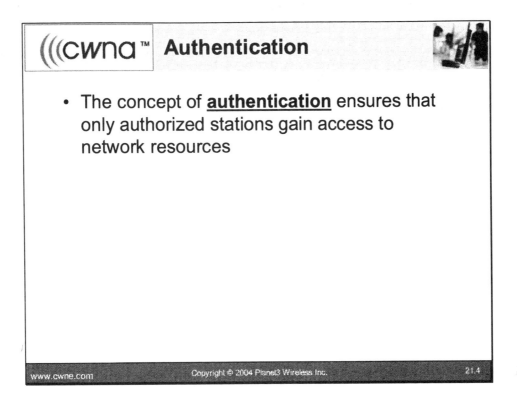

The concept of authentication ensures that only authorized stations gain access to network resources. A station realizes that authentication is required. It sends an authentication frame to the station with which it is attempting to authenticate. This frame is always sent unencrypted. The station identifies the authentication algorithm being used. The authentication frame contains the identity of the station being authenticated and information specific to the algorithm being used. A sequence of authentication frames is exchanged, the specifics of which will vary depending on the specifics of the authentication algorithm in use. These frames may or may not be encrypted. The final frame contains the result of the authentication (successful or unsuccessful).

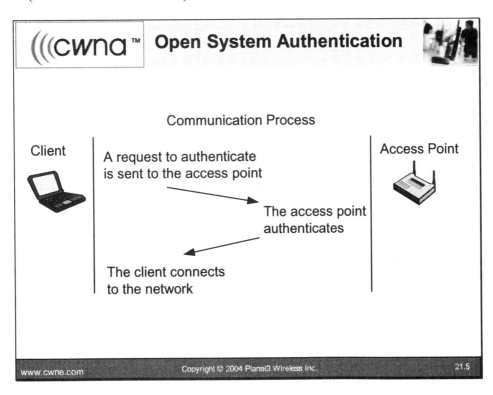

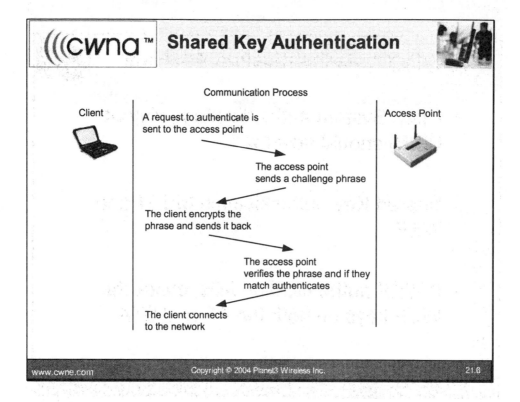

When using Shared Key Authentication, the WEP key is used for authentication and then subsequently used for encrypting data frames.

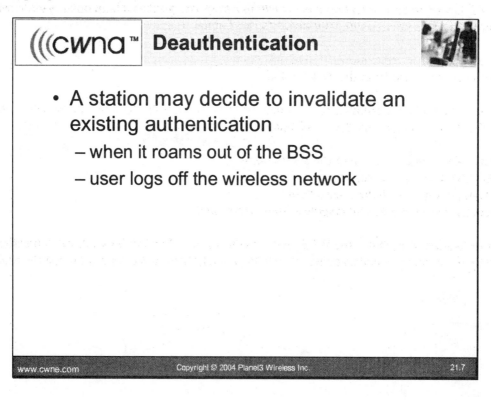

A station may decide to invalidate an existing authentication (e.g. when it roams out of the BSS or the user logs off the wireless network). The concept of deauthentication provides for this functionality. To deauthenticate, the station sends a deauthentication message to the access point from which it is deauthenticating. The deauthentication message contains the address of the deauthenticating station, and contains the address of the access point with which authentication exists. No response is expected from the access point.

Authentication Analysis

- Open System authentication (without WEP) should never fail

- Shared Key authentication <u>MUST</u> use WEP

- If WEP authentication fails, check the WEP keys on both the AP and STA

If Open System authentication is being used, authentication should never fail (unless the manufacturer is using the WEP key as an authentication mechanism in a proprietary manner). In open systems without WEP, the authentication process typically involves two frames:

1. Open System Authentication to the access point
2. Successful response from the access point

In open systems with WEP, mismatching keys may be used by mistake. Shared Key authentication MUST use WEP and typically involves four frames:

1. Shared Key Authentication to the access point
2. Challenge from the access point
3. Challenge response to the access point
4. Successful or Unsuccessful response from access point

If WEP authentication fails, check the WEP key(s) in the station. The last frame in the authentication conversation will contain a response code, which can be interpreted to see the reason that the station was rejected.

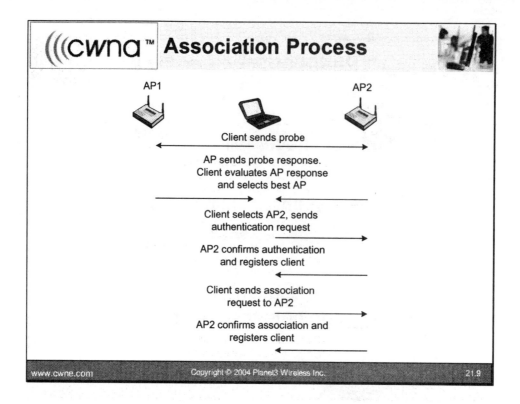

To deliver a message within a distribution system, the DS needs to know which access point in the DS is capable of reaching the destination station. Association provides this information. A station discovers that an access point is within its coverage area. The station sends an association request frame to the access point. This association request contains the MAC address of the station, the MAC address of the access point, and the ID of the ESS that the station is joining (called an ESSID or just SSID).

The access point determines whether the station may join the BSS and sends an association response. The association response contains the result of the requested association (successful or unsuccessful). Association is sufficient to ensure communication in all cases where stations remain within a single BSS.

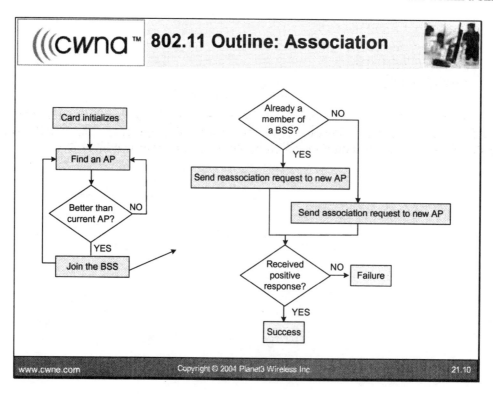

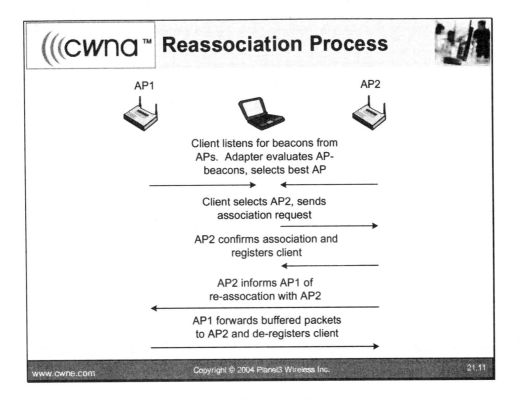

When a station moves from the coverage area of one BSS to another, it may be best for performance to associate with the new BSS. Reassociation provides for this circumstance, and utilizes the following process:

1. The station realizes that another access point is preferable to the one with which it is currently associated
2. The station sends a reassociation request frame to the preferable access point. The reassociation request contains:

   - the MAC address of the station
   - the MAC address of the new access point
   - the MAC address of the currently associated access point
   - the ID of the ESS

The new access point sends a reassociation response to the station. The reassociation response contains the result of the requested reassociation (successful or unsuccessful). Reassociation is sufficient to ensure communication in cases where a station moves from one BSS to another within an ESS.

Reassociation and the Distribution System (DS)

- APs do not typically update switches

- APs often update APs in a proprietary manner

- Some manufacturers implement no reassociation (roaming) process

A station can only be associated with one access point at a time. When a station reassociates, something must happen within the DS to ensure that the ESS realizes that the station has moved to a different BSS. Specifically, the old access point must know that the station is no longer associated with it. Notice that the reassociation request/response sequence does not directly accomplish this. As stated earlier, the 802.11 specification does not describe the details of the distribution system. Correspondingly, the means by which the DS handles reassociation is not described by the 802.11 specification. Vendors currently implement this proprietarily, which results in incompatibility between different vendors' access points.

Roaming Process

- Stations initiate a move from AP to AP due to:
  - STA moving away from current AP
  - Interference on the current channel
  - Current AP has ceased operation
  - Channel of the current AP is busier than alternative available channels

Roaming involves switching of a station to another access point as result of a station moving away from the current access point. Roaming can be caused by:

- Interference on the current channel
- The current access point has ceased operation
- The channel of the current the is busier than alternative available channels

The initiative for changing to another access point is taken by the station, based on a proprietary mechanism determined by the manufacturer. For example, Proxim's Orinoco uses "Combined Communications-Quality & Load" (CCQL). CCQL is determined by the following factors:

- SNR (Signal to Noise Ratio) on path with current access point using a running Average Signal Level from beacon receptions and a running Average Noise Level from all receptions in current channel
- Load on the channel
- Result of Sweep (when Searching) and the SNR of probe responses

Roaming Process

- Manufacturers may only use one or more criteria such as:

  – Signal Strength

  – Bit Error Rate (BER)

  – Signal-to-Noise Ratio (SNR)

  – Load

Manufacturers may use one or more criteria for initiating roaming, such as:

- Signal Strength
- Bit Error Rate (BER)
- Signal-to-Noise Ratio (SNR)
- Load

Some manufacturers want their radio to roam upon finding a better access point, and some want their radio to wait until a minimum signal quality threshold is reached first. How stations are designed to roam determines the uses for which they are best designed.

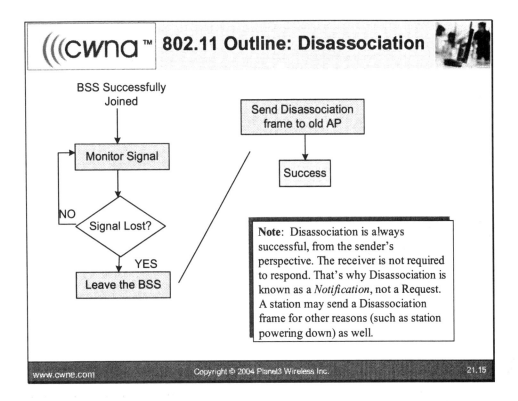

When a station leaves a BSS, it may choose to terminate its association with the BSS. Disassociation provides this functionality. The station sends a disassociation notification to the access point with which it is currently associated. The disassociation notification contains the MAC address of the station, and the MAC address of the access point with which it is associated. No response is expected.

When an access point leaves the BSS (e.g. because it powers down), it will also send a disassociation notification, which contains the broadcast address as the address of the station and the access point's MAC address as the associated access point. Although stations are required to send disassociation notifications when they leave the network, the 802.11 MAC layer does not depend on a station doing this.

Association/Disassociation Analysis

- The most common reason for failed association:

  – The device is not authenticated

  – The device is prohibited from associating by a MAC filter

  – Incompatible data rates

In general, association will succeed. If association does not succeed, the association response frame will contain a code that explains the reason why. Wireless packet analyzers can interpret these codes. The most common reasons for association to fail are:

- The device is not authenticated
- The device is prohibited from associating by a MAC access list

Another common reason for failure is incompatible data rates. The station may not support the proper data rates to send and receive data through the access point. Examine the Supported Data Rates element in the frame. Disassociation frames will contain a code explaining the reason why the station is disassociating or being disassociated:

- Station idle too long
- Access point is overloaded
- Station going out of range

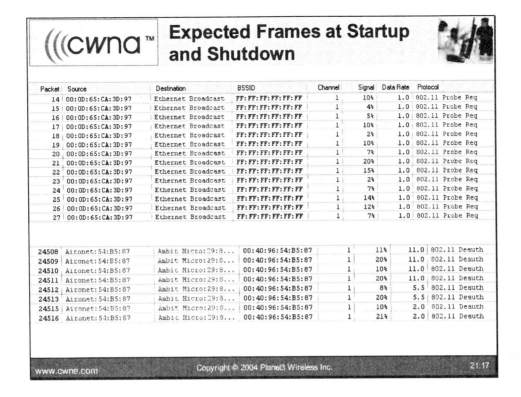

When a wireless station starts up or first joins a BSS, these frames will usually be seen in the following order:

1. Probes / Probe Responses - Optional, since a station may just listen for beacons; Station will listen for probe responses and choose an access point
2. Authentication - The station authenticates with the access point. This will occur even in Open Systems.
3. Association / Association Response - The station associates with the access point
4. Data
5. Disassociation - This frame may not be seen, depending on how the station leaves.
6. Deauthentication - This frame may not be seen, depending on how the station leaves.

Chapter 22
Power over Ethernet (PoE) and 802.3af

Objectives

Upon completion of this chapter you will be able to:

- Explain Power over Ethernet (PoE) and what role it plays in enterprise wireless LANs
- Explain end-span and mid-span technology as defined by the 802.3af standard
- Describe the various PoE adapter types and how they work

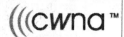
Proprietary vs. Standards-based

- Many manufacturers use non-standard pins and voltages

- Most manufacturers are standardizing, <u>but not all</u>

- Some wireless LAN switch and middleware vendors are implementing switchable -/+ 48 VDC

Many manufacturers use non-standard pins and voltages such as -24VDC or -48VDC since the 802.3af standard has only recently been ratified. Not all manufacturers are rushing to standardize on pins and voltages due to large initial investments in proprietary PoE hardware. Some wireless LAN switch and middleware vendors are implementing switchable -/+ 48VDC systems for broader compatibility with access point vendors.

IEEE 802.3af - Power over Ethernet (PoE) standard

- End-span
  - PoE in the switch

- Mid-span
  - Single-port injectors
  - Multi-port injectors (patch panels)

- Voltage = +48 VDC

The IEEE 802.3af standard defines two types of power source equipment:

1. End-span refers to an Ethernet switch with embedded Power over Ethernet technology. These switches may deliver data and power over the same wiring pairs - transmission pairs 1/2 and 3/6 or over spare pairs 4/5 and 7/8.

2. Mid-span devices resemble patch panels and typically have between six and 24 channels. They are placed between legacy switches and the powered devices. Each of the mid-span ports has an RJ-45 data input and data/power RJ-45 output connector. Mid-span devices tap the unused wire pairs 4/5 and 7/8 to carry power, while data runs on the other wire pairs.

For new deployments, consider buying an end-span Ethernet switch. Mid-span panels make sense for upgrading a network without replacing switches and for low port density. It is wise to consider deploying a new end-span switch because it will be attached to IP phones, wireless LAN access points, and other popular powered terminals during its expected life span.

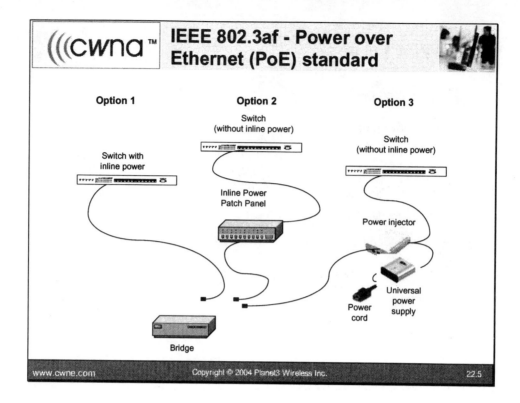

+48 VDC is typically delivered to each device, but the standard says +44VDC to +57VDC is acceptable. Power to a single node is limited to 12.95 W. Current to a single node is limited to 350 mA. IP Phones and access points typically take 3.5-10 Watts to operate. Only terminals that present an authenticated PoE signature will receive power, preventing damage to legacy equipment.

PoE Works with existing cable plant, including:

- Category 3, 5, 5e or 6
- Horizontal and patch cables
- Patch-panels
- Outlets

Some PoE switches cannot power PoE devices on every port due to a lack of overall power.

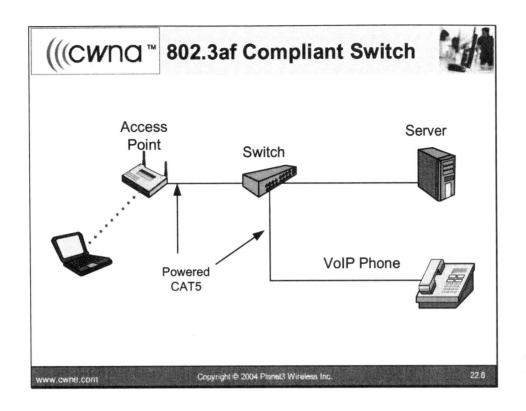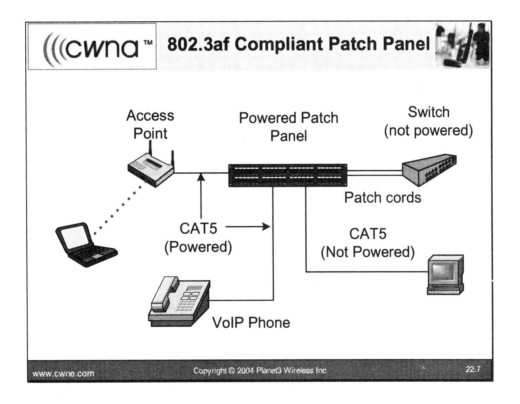

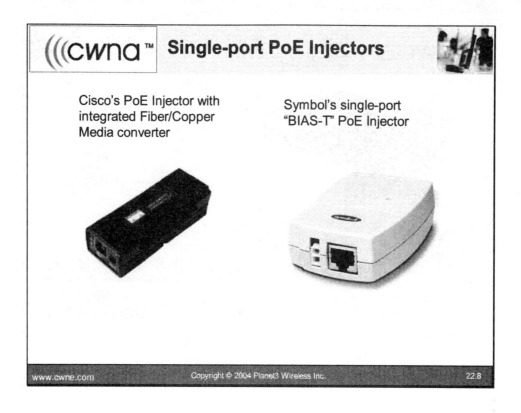
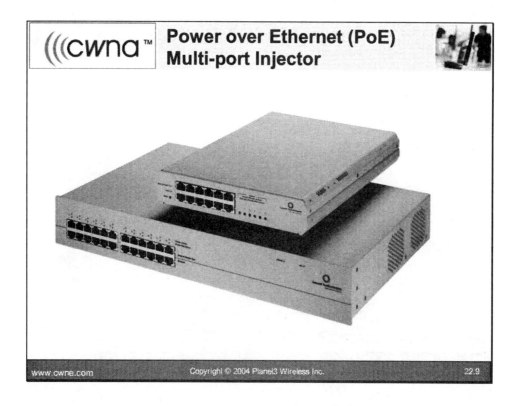

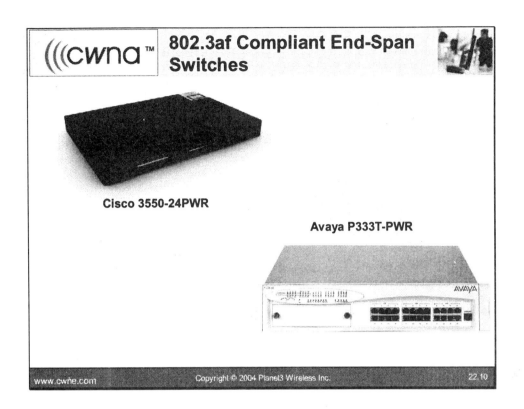

Chapter 23
Fragmentation and Acknowledgements

Objectives

Upon completion of this chapter you will be able to:

- Explain how and why acknowledgements are implemented according to the 802.11 standard
- Explain how frame fragmentation is implemented according to the 802.11 standard

 **Acknowledgements**

- In 802.11, the Acknowledgement for a frame <u>must immediately follow</u> the frame which is being acknowledged

- No one else within hearing range of the transmitting station is allowed to send data until the acknowledgement occurs

In 802.11, the acknowledgement for a frame must immediately follow the frame which is being acknowledged. This means that 802.11 is a "ping-pong" style reliable protocol. The sequence number in transmitted frames is used to reconstruct fragmented frames, but is not used at all in the acknowledgement process. No one else within hearing range of the transmitting station is allowed to send data until the acknowledgement occurs. Of course, 802.11 engineering ensures that the ACK comes back very quickly, in order to avoid performance slowdowns while waiting.

Which frames are acknowledged

- All unicast frames are acknowledged

- Multicast and broadcast frames are never acknowledged

All unicast frames are acknowledged:

- Unfragmented data frames
- Fragments of a data frame
- Directed management frames
- Association Request
- Authentication Request

Multicast and broadcast frames are never acknowledged:

- Beacon frames
- Broadcast data frames

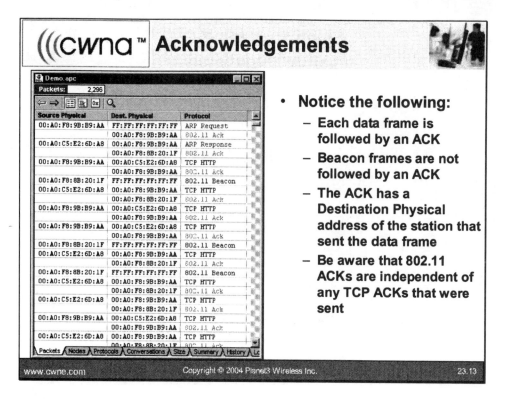

802.11 retransmissions (transmission retries) occur when a station's packets are corrupted or lost. If a single station has a high incidence of retransmissions, check for weak signal strength at that station and/or interference from other RF transmitters near that station.

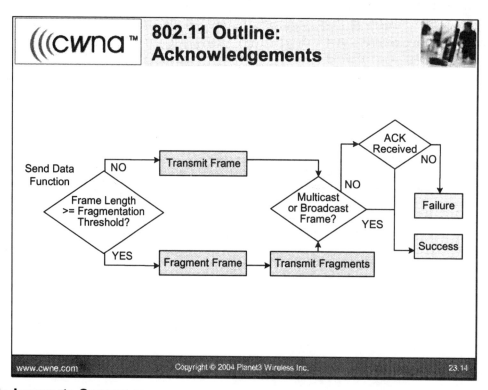

Acknowledgements Summary

Data frames and frame fragments are individually acknowledged. If the frame is a multicast or broadcast frame, no MAC layer acknowledgement is required, and the transmission is always considered successful. If the frame is a unicast frame, acknowledgements must be received (a single acknowledgement for each frame or fragment). If all acknowledgements are received, the transmission is a success. If not, it is a failure, and the frame must be retransmitted.

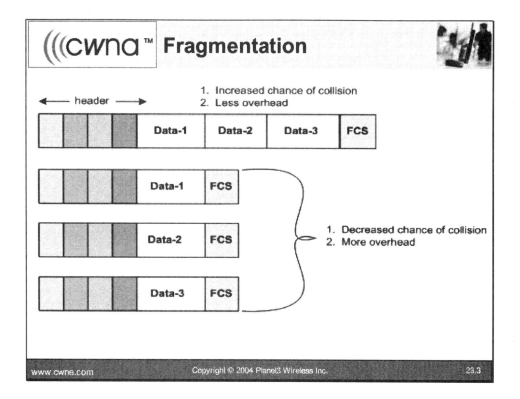

Packet fragmentation is the process by which a block of data is broken down into smaller blocks of data before transmission. In traditional wired LANs, packet fragmentation is rarely used. Low corruption rates mean that it is more efficient to send a few large packets than to send many small packets. Many small packets actually increase overhead. Fragmentation is only required when extremely large blocks of data (larger than the maximum packet size) must be moved. Fragmentation is essential in a wireless environment and is built into the MAC layer. Much higher corruption rates mean that it may be more efficient to send many small packets than to send a few large packets. This minimizes the negative effects of a corrupted packet which requires retransmission, by only have to retransmit a few bytes, instead of a lot.

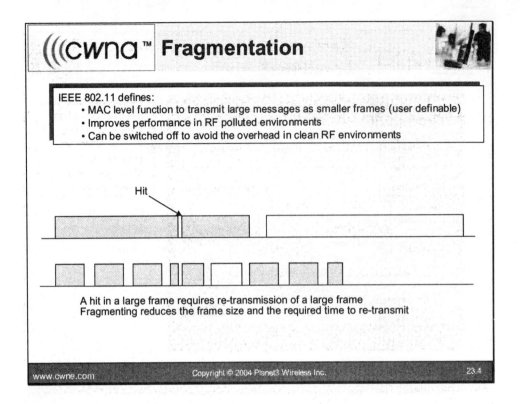

Fragmentation will occur any time the MAC layer receives a block of data with a length greater than the Fragmentation Threshold. This threshold is configured in the station or AP. The maximum 802.11 frame size is 2346 bytes, but the network layer limits packets to 1500 byte payloads. The administrator should configure this threshold for optimal performance. Thresholds that are too small result in unnecessary fragmentation. Thresholds that are too large result in too much data being retransmitted. The 802.11 standard defines a reliable MAC layer. Fragments (and other 802.11 MAC frames) are acknowledged. If an acknowledgement is not received within a certain time period after transmission, the frame is retransmitted. A reliable MAC layer is quite unusual in the LAN world. The novel demands of wireless networking require it.

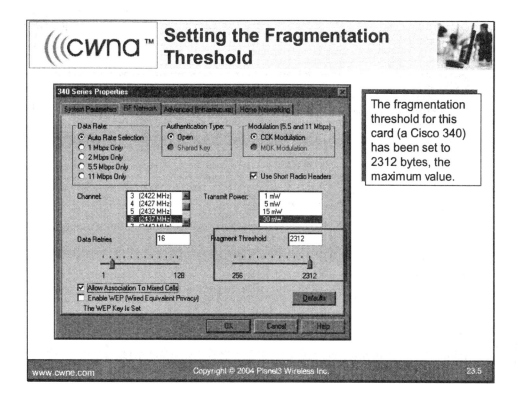

Fragmentation Process

A block of data, larger than the fragmentation threshold, is received by the MAC layer. The MAC layer breaks the block down into several fragments. All fragments except the last are equal in size to the fragmentation threshold. The last fragment has however many bytes are left over. Each fragment is given its own MAC header and CRC. Each fragment is transmitted as a separate frame. The next station to receive the fragments acknowledges them. If an acknowledgement is not received for a given fragment, the fragment will be retransmitted. When the entire frame is received and reassembled, the receiving station passes it up the protocol stack. Multicast and Broadcast frames are never fragmented.

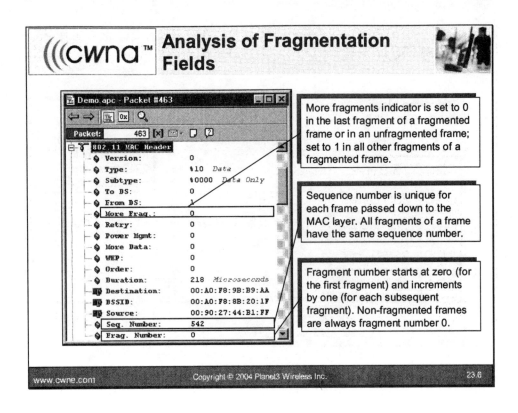

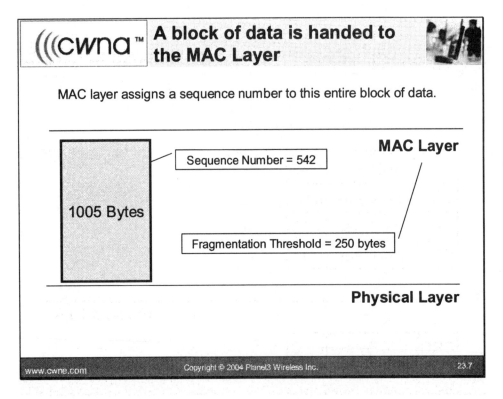

The Block is Fragmented

Since the block is larger than the Fragmentation Threshold, the MAC layer will fragment it

CRC	5 Bytes	MAC	Seq. # = 542; Fragment # = 4; More Frag. = 0
CRC	250 Bytes	MAC	Seq. # = 542; Fragment # = 3; More Frag. = 1
CRC	250 Bytes	MAC	Seq. # = 542; Fragment # = 2; More Frag. = 1
CRC	250 Bytes	MAC	Seq. # = 542; Fragment # = 1; More Frag. = 1
CRC	250 Bytes	MAC	Seq. # = 542; Fragment # = 0; More Frag. = 1

Physical Layer

Each Fragment is Transmitted

- The receiver will reconstruct the original block of data
- The receiver will acknowledge each fragment individually

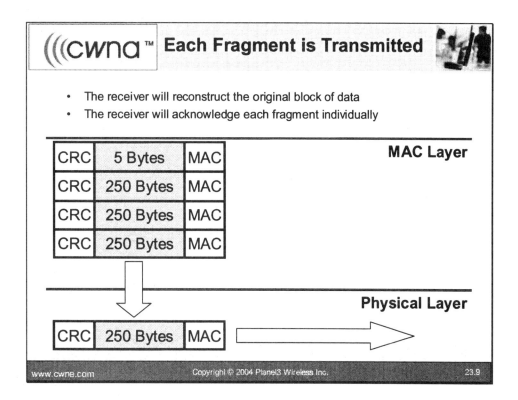

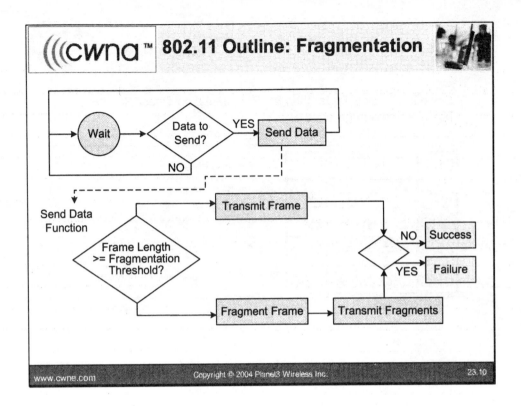

The fragmentation threshold should be set based on the percentage of corrupted frames in the environment. There is no rule for what is appropriate. In general, more corrupted frames will lead to a lower "best" fragmentation threshold. Throughput testing with various fragmentation thresholds should be performed to determine the optimum value for the environment. Each fragment of a frame is acknowledged individually. Fragments are transmitted in a "burst". All of the fragments of a frame should be transmitted and acknowledged before any other frame is sent.

Chapter 24
Standards-Based Roaming

Objectives

Upon completion of this chapter you will be able to:

- Understand that vendors purposefully implement roaming differently
- Understand the basic functionality of the 802.11f (IAPP) recommended practice
- Understand how routed boundaries affect wireless LAN roaming
- Understand why MobileIP is used in a wireless environment

Roaming in an ESS

- Roaming is usually a key component of WLANs

- The 802.11 standard does not dictate the specifics of roaming within an ESS

- 15-20% overlap necessary

- Roaming based on RSSI values

- Roaming mechanisms are usually proprietary

Being able to roam between access points to maintain wireless connectivity is one of the key advantages of wireless LANs. The 802.11 standard does not dictate the specifics of roaming within an ESS (between BSAs). Some vendors implement proprietary L2 mechanisms for roaming. Some vendors implement 802.11f IAPP. Some vendors have no L2 mechanism for roaming at all.

15-20% of cell overlap is generally required to roam gracefully between access points. Roaming between access points from the same vendor is usually much faster than between different vendors.

Some vendors cater to vertical markets that require mobile connectivity (seamless connections while moving). Some vendors cater to vertical markets that need only non-mobile wireless connectivity (connecting in multiple places but not needing connectivity while moving between them). Roaming thresholds (often based on an arbitrarily cumulative value called RSSI) are set differently by each manufacturer.

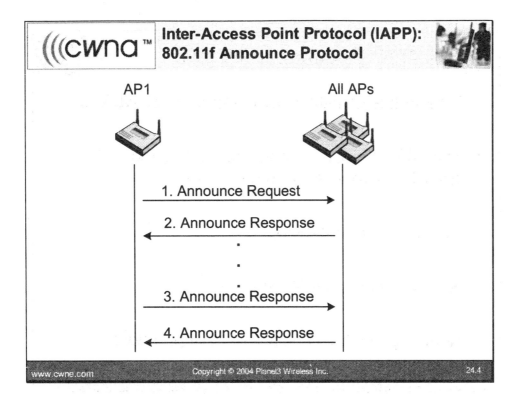

802.11f is called "Recommended Practices for Multi-Vendor Access Point Interoperability via Inter-Access Point Protocol Across Distribution Systems Supporting IEEE 802.11 Operations". 802.11f adds definitions for access points and distribution systems absent from the IEEE 802.11 standard. 802.11f is NOT a standard but may serve the same purpose as a standard.

802.11f specifies a protocol most often referred to as the Inter-Access Point Protocol (IAPP).

The IAPP Process

1. At startup, the access point transmits an "announce request" (IP Multicast Destination Address) using defined UDP/IP group addressing.
2. Access points that are part of the same network and are already operational will respond with an "announce response", containing the IP address of the replying access point and the BSSID of the replying access point.
3. The new access point uses the data in the reply to build a BSSID-to-IP conversion table to relate the BSSID, (used by the roaming station to identify its "old" access point) to the IP address of the "old" access point. Future implementations will carry items such as the name of the access point, to be used to identify access points in the "Site Surveying" displays, and authorization information regarding the mobile station in case a centrally based authentication scheme is used.
4. After an appropriate time interval, when all responses are received, the "new" access point will issue an "announce response" to indicate its operational status. The "new" access point will (as will all access points) re-issue the "announce response" to keep informing all participating access points about any changes in the status.

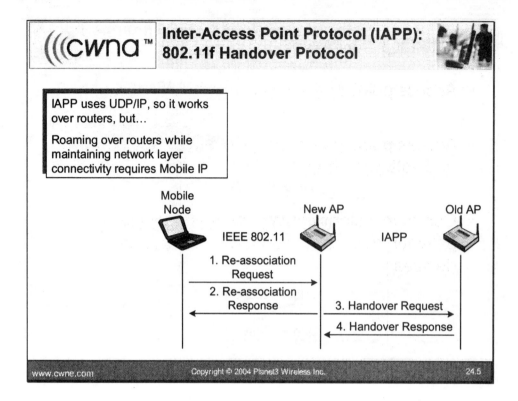

1. When the mobile station moves away from its "old" access point, it issues a re-associate request to a "new" access point.
2. The "new" access point will return a re-association response when it accepts the roaming station. The access point service for the mobile station starts at this point in time.
3. The "new" access point sends a hand-over request to the old access point (via the Distribution System). The IP address of the old access point is determined based on the BSSID carried in the re-association request.
4. When the hand-over response received, the hand-over is considered to be completed.

When STA is operating under IEEE 802.1x security, hand-over response will contain authentication information concerning the station. Account information from RADIUS server concerning the station is cached to avoid having to re-initiate authentication.

LWAPP

- Access point <u>device discovery & authentication</u>

- Access point <u>information exchange, configuration, and software control</u>

- Communications <u>control and management</u> between access point and wireless system devices

---

One of the problems with 802.11f and proprietary Layer2 hand-off mechanisms is updating of the distribution system (DS). The access points and Ethernet switches do not share a common language. Access points do not notify the directly-connected switch that a user has roamed to or from it. A recent IETF draft has been set in motion detailing a new protocol deemed "Lightweight Access Point Protocol" (LWAPP). LWAPP is likely to become a subset of the IETF standard called "Control and Provisioning of Wireless Access Points" (CAPWAP).

LWAPP will specify:

- Access point device discovery & authentication
- Access point information exchange, configuration, and software control
- Communications control and management between access point and wireless system devices

The final draft of LWAPP is expected in 2005.

Chapter 25
Bandwidth Control, Network Management, & AAA

Objectives

Upon completion of this chapter you will be able to:

- Explain the need for bandwidth control devices in a wireless LAN and how they work
- Explain the need for network management hardware and software and their role in a wireless LAN
- Explain the need for AAA in a wireless LAN and how it can be implemented

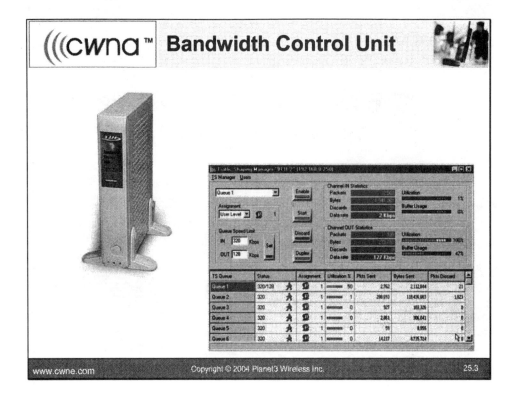

Bandwidth control units are generally used by wireless ISPs and assign bandwidth filters based on MAC addresses and queues. Many of today's routers, switches, and gateways have integrated bandwidth controls and are used for enterprise indoor installations.

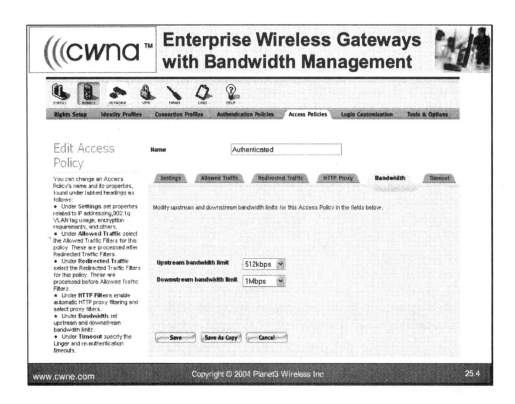

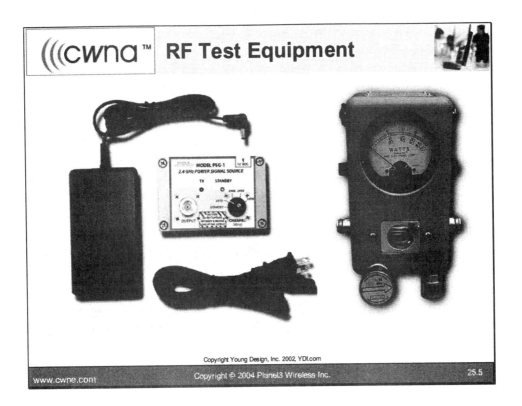

Any number of test equipment devices may be used by an administrator for testing proper functionality of the equipment. A power meter and signal source may be used to test coaxial cables, antennas, connectors, and other devices for water damage, intermittent connections, etc. A spectrum analyzer may be used to see if a transmitter is emitting the correct amount of output power or too much background noise. A spectrum analyzer is used to find RF interference sources and to check the ambient RF noise in an environment.

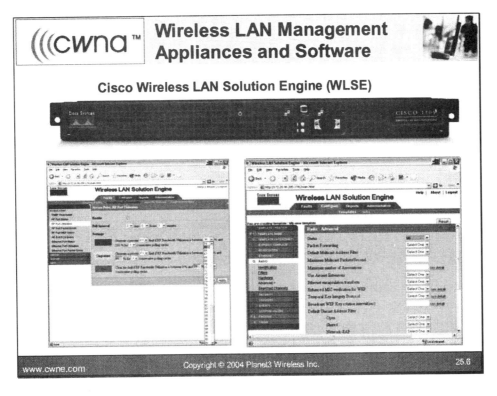

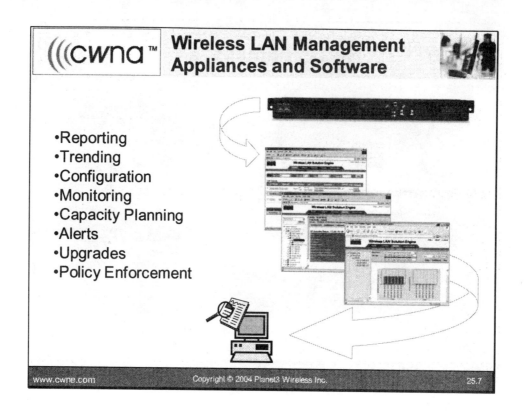

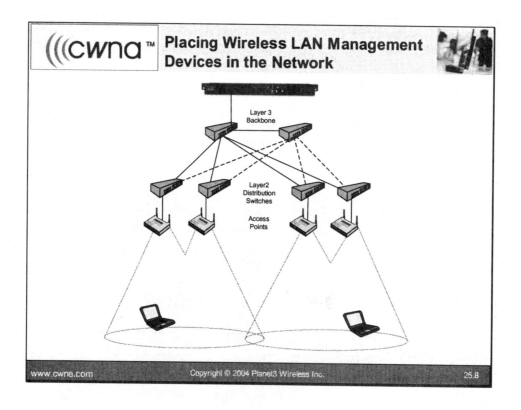

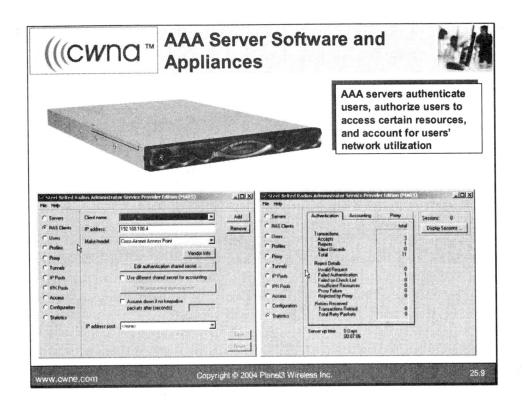

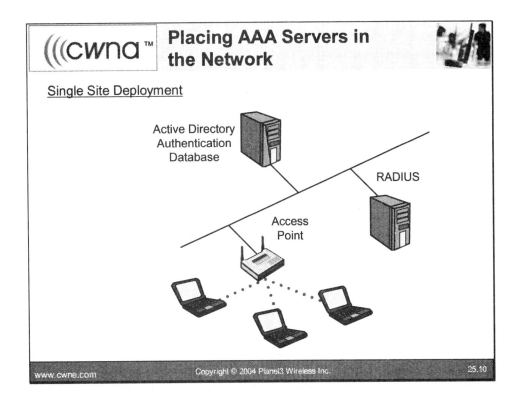

In simple deployments such as this, RADIUS is typically used to proxy authentication requests to an existing centralized user database such as NDS, Active Directory, LDAP, or others.

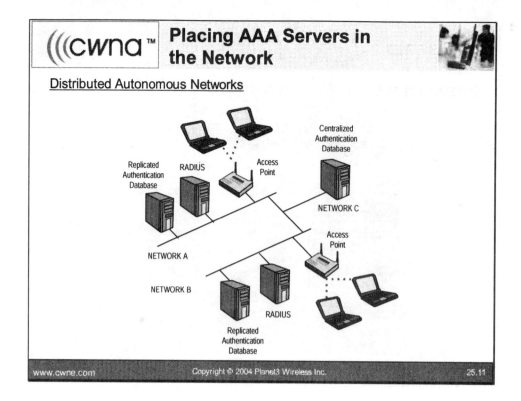

In distributed autonomous networks, the authentication database is replicated from the central site to each autonomous site. All authentications occur locally.

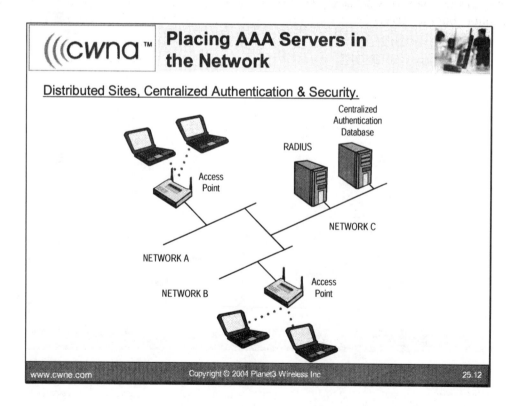

In a network such as this, users often send their credentials across a WAN link to a central IT data center. Such a solution is very cost effective. However, often this configuration is very slow, causing network connectivity problems locally due to congested WAN links. Often, the entire authentication process happens over slow links.

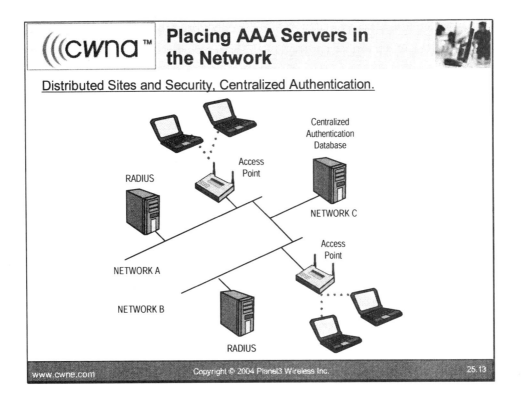

This type of AAA design can be expensive due to local authentication services at each location. This design has the advantages of simple management and fast client connectivity. Even if slow WAN links are used between sites, only very small amounts of data are sent over the link. Local AAA servers proxy the authentication, but the servers handle most of the authentication process locally.

Chapter 26
Routers, Gateways, Switches, and VoWiFi

Objectives

Upon completion of this chapter you will be able to:

- Understand the role of a wireless router in a network and how it works
- Understand the role of an enterprise wireless gateway in a network and how it works
- Understand the role of a wireless LAN switch in a network and how it works
- Understand the role of a wireless LAN antenna switch in a network and how it works
- Describe the components of a simple VoIP system operating over a WLAN
- Describe problems associated with VoIP over a WLAN

Wireless LAN Routers

- Wi-Fi Compliant AP
- L2 Security (WEP, 802.1x/EAP)
- L3 Security (VPN Client & Server) - L2TP, IPSec, PPTP
- Firewall
- Interface Bridging
- IP Routing with Static & RIPv2 routes
- VPN Filtering
- Firewall features

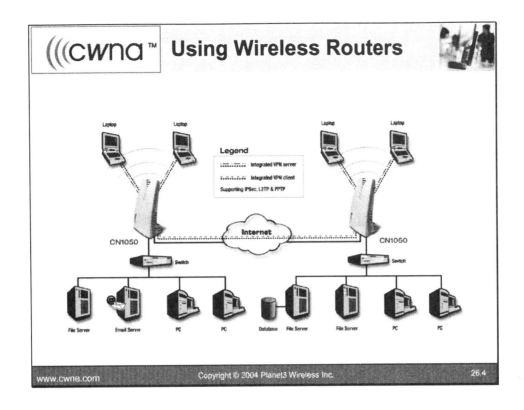

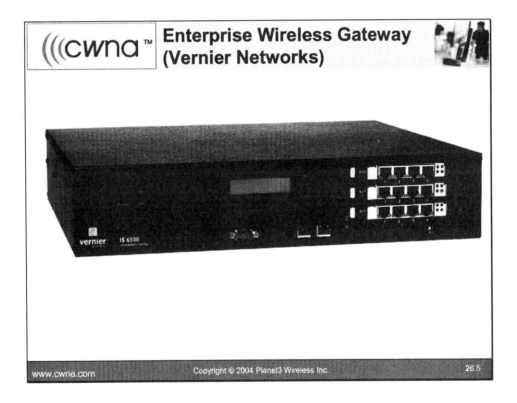

- Routing / Bridging / Filtering features
- Role-based Access Controls by location, time, and ID
- PoE for APs, DHCP Server, LDAP database support
- CLI, HTTPS, Telnet, SNMP management
- VLANs, VPNs, 802.1x/EAP, L3 Roaming

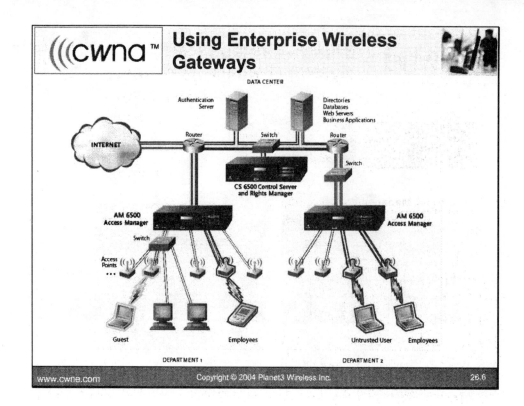

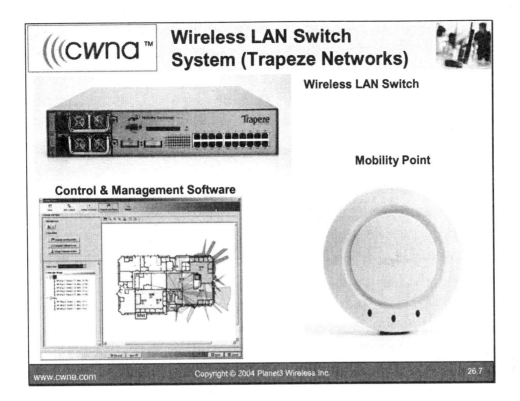

Access Points

- Sometimes have vendor-specific names such as access ports or mobility points
- Alternate naming intended to differentiate between standard "thick", "smart" or "fat" access points and new variations deemed "thin" (having minimal functionality) or "integrated" (having optimized functionality)
- Vendors' philosophies on access point architecture vary between extremes
- Access points should be very intelligent
- Access points should be optimized for the environment
- Access points should be dumb

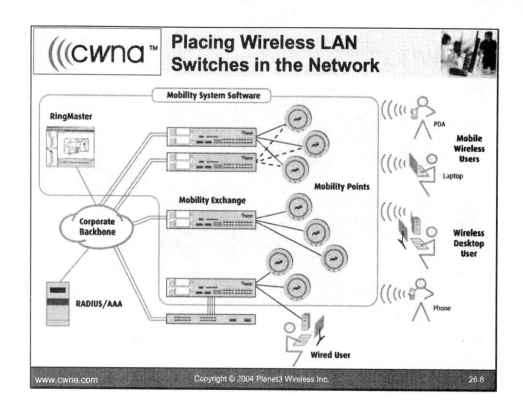

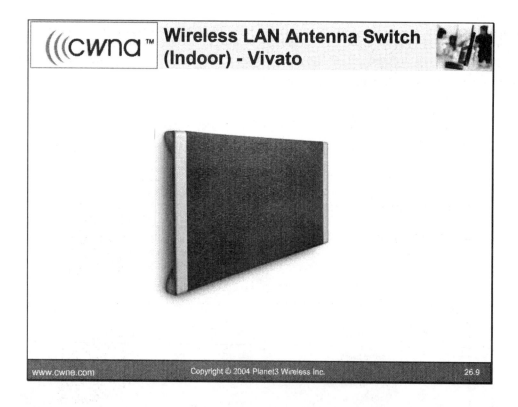

Vivato's switches use phased-array radio antennas to create highly-directed, narrow Wi-Fi beams. The Wi-Fi beams are created on a packet-by-packet basis. Vivato calls this technology PacketSteering™. Unlike current wireless LAN broadcasting, Vivato's switched beam is focused in a controlled pattern and pointed precisely at the desired client device. These narrow beams of Wi-Fi enable simultaneous Wi-Fi transmissions to many devices in different directions, thus enabling parallel operations to many users.

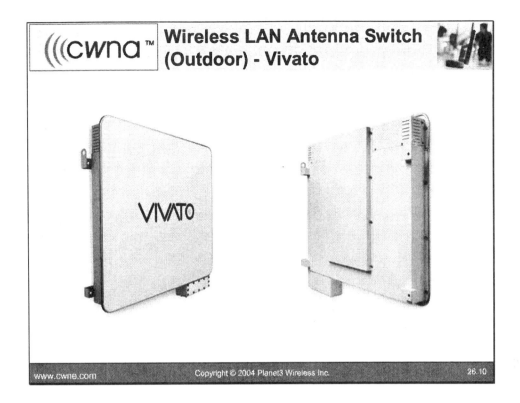

These narrow beams reduce co-channel interference, since they are powered only when needed. Vivato's Wi-Fi switches significantly increase the range of Wi-Fi. Rather than transmit the radio energy in all directions, Vivato's PacketSteering™ concentrates the same amount of energy into a narrow, long beam. A beam is effectively a high-gain antenna that is formed for the duration of a packet transmission. The result is extreme range, extending the reach of Wi-Fi from tens of meters to kilometers.

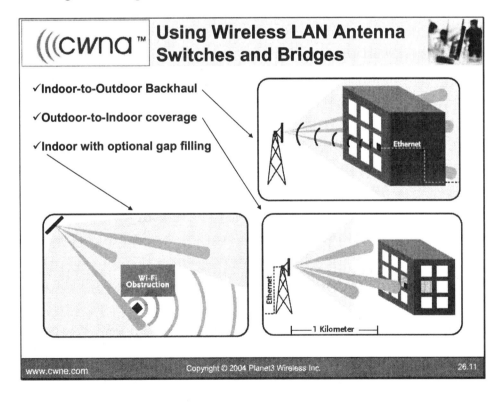

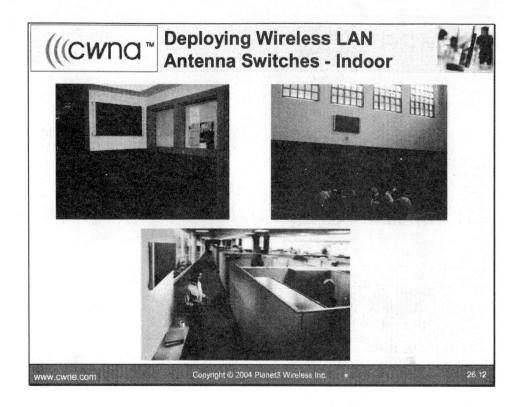

Indoor panels are used for covering entire floors using both front and rear lobes of the phased antenna array. Floors above and below are covered to some degree as well.

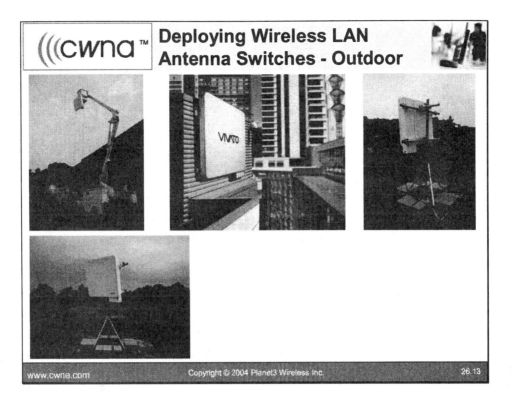

Outdoor panels are typically mounted on buildings adjacent to the building that requires coverage. Because each panel weighs approximately 80 pounds, bucket trucks are often used during deployment to place panels near potential mounting locations. Notice the downtilt of at least 4 degrees to maximize ground coverage.

Electrical runs and fiber connections are permanently installed after placement is final. Fiber is often used instead of Cat5 as an added lightning protection mechanism. Due to the size/weight of the outdoor panels, concrete blocks are often used for holding mounting units in place. High winds may move the unit because it is not perforated like a grid antenna would be. Indoor and outdoor units are identical. Outdoor units have the indoor units placed in a radome NEMA enclosure with a heat exchanger and cooling fans. Radomes provide excellent weather protection and can be painted.

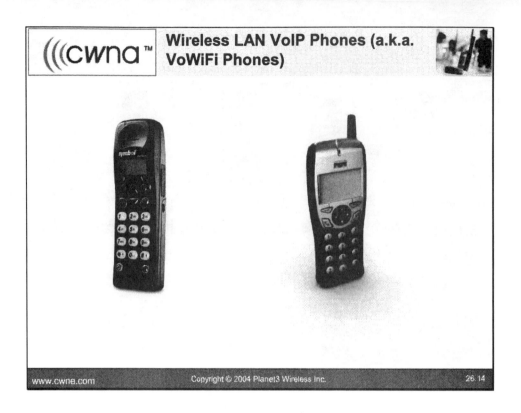

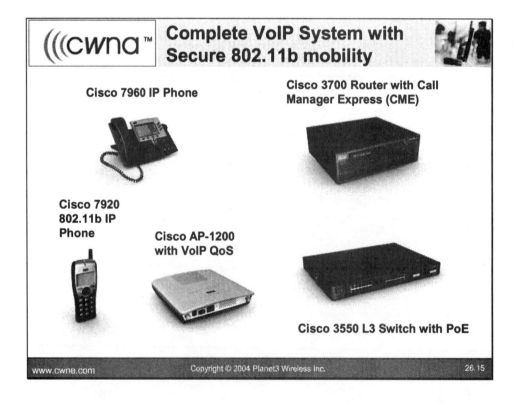

Many vendors have solved the subnet boundary problem with MobileIP, Proxy MobileIP, DAT, and other similar solutions. 802.11f (IAPP) will solve some intra-subnet roaming problems. 802.1x/EAP and VPN solutions solve security issues over most every type of wireless LAN link. Many vendors have proprietary solutions for voice and video QoS and 802.11e is on the way. None of these solutions are fast enough to give a reasonable QoS to users on the move while maintaining security.

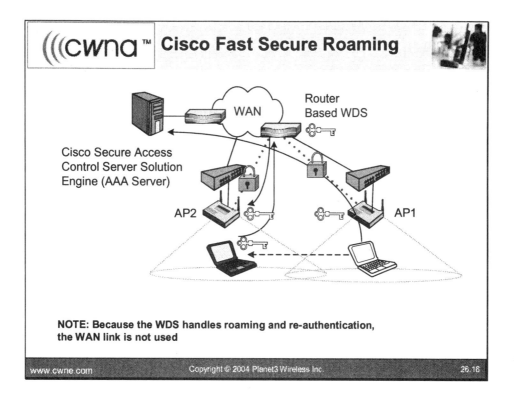

Fast Secure Roaming (FSR) is a Cisco proprietary protocol that allows authenticated client devices to roam securely from one access point to another without any perceptible delay during re-association. FSR supports latency-sensitive applications such as VoWiFi, Enterprise Resource Planning (ERP), or Citrix-based thin client solutions, without dropping connections during roaming.

Cisco's additional router and switch code that supports mobile networking, deemed WDS, provides fast, secure handoff services to access points for <150ms roaming within a subnet. Cisco FSR requires Cisco or Cisco compatible client devices that support the Cisco Centralized Key Management (CCKM) protocol.

The FSR Process

- Access point must now use 802.1x to authenticate with the WDS access point (AP1) to establish a secure session
- Initial client 802.1x authentication goes to a central AAA server (~500ms)
- During a client roam, the client signals to the WDS that it has roamed and WDS will send the client's key to the new Access Point (AP2)
- The overall roam time is reduced to <150ms, and in most cases, <100ms

Chapter 27
Power Management

 Objectives

Upon completion of this chapter you will be able to:

- Explain the need for power management in a wireless LAN environment
- Explain the difference between CAM and PSP modes
- Explain how the 802.11 standard implements both CAM and PSP modes in Infrastructure and Ad Hoc networks

Power Management Modes

- Continuous aware mode (CAM)
  - the mode where no power-saving features are enabled
- Power save polling mode (PSP)
  - power saving mode defined by the 802.11 standard that allows stations to save power by being powered down (sleeping) while inactive on the network while at the same time allowing them to awake to receive packets destined to them

Typical Power Draw

Data-Rate	Peak power consumption		
	TX	RX	Doze
11 Mbps	285 mA	185 mA	9 mA
5.5 Mbps	285 mA	185 mA	9 mA
2 Mbps	285 mA	185 mA	9 mA
1 Mbps	285 mA	185 mA	9 mA

Wireless devices often run on batteries, so they may have limited battery life. The 802.11 standard offers a power management mode to conserve battery life. When a device operates in power management mode, the access point does not immediately forward frames to the device. Devices can put their NIC into "sleep" mode to save power. The access point buffers frames for the device until a pre-determined time. Beacons are used as time-intervals. At that time, the device wakes up its NIC, receives and sends frames, and then puts the NIC back to sleep. The power management field indicates whether a device is operating in power management mode or not.

Power management will generally result in lower throughput because a station can't receive data immediately. The station has to wait for the access point to send it data at the "wake up" time. The station has to notify access point that it is awake and willing to receive traffic. The station can send data whenever it wants, but must power up the NIC before transmission. There will be some delay as the card re-synchronizes itself with the rest of the network. Power management can decrease power consumption by a factor of hundreds. Actual power savings will vary, depending on the environment and the usage patterns of the clients.

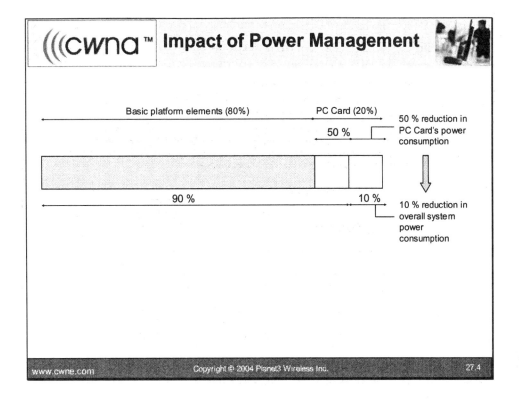

A common misconception regarding systems not designed for low power consumption (such as a laptop computer) is that power management features will make your battery last 2-3 hours longer.

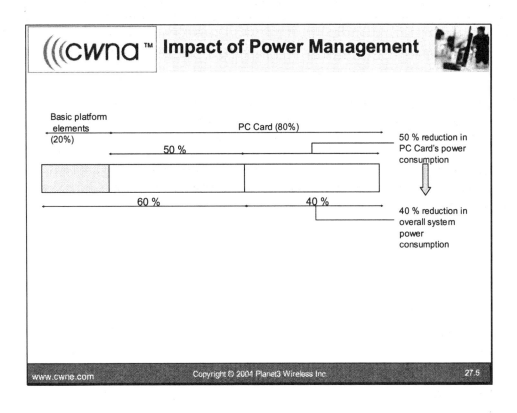

In a platform that is designed for low power consumption, e.g., no back-light on screen, no rotating media (hard drives & CD-ROMs), and a low power processor; you can expect power management features to be quite effective.

Power affects throughput

- Throughput measurements on notebook computer
- Large file (7 MB transmission)

	Network disk to Notebook	Notebook to network disk	Average Battery life
With Power Management	213 sec	422 sec	128 minutes
Without Power Management	62 sec	89 sec	102 minutes

Applicability of Power Management Features

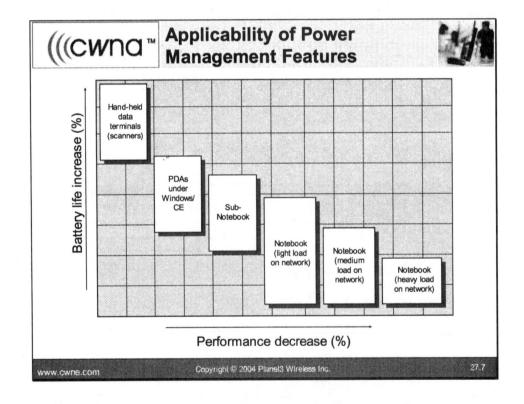

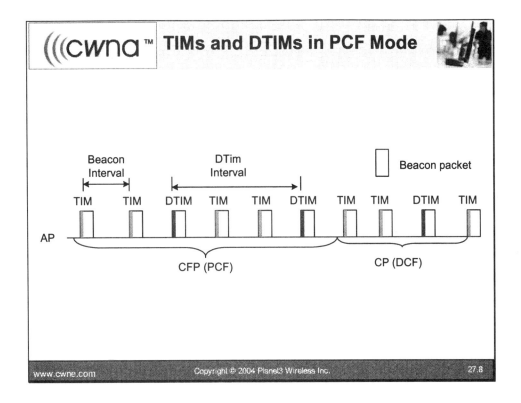

Beacons contain TIMs (Traffic Indication Maps) and DTIMs (Delivery Traffic Indication Messages), which are used for notifying stations that have just awakened that they have traffic queued at the access point. The DTIM is often sent every 2nd or 3rd beacon by default (configurable). DTIMs are used for notifying recently awakened stations that broadcast and multicast traffic is queued at the access point.

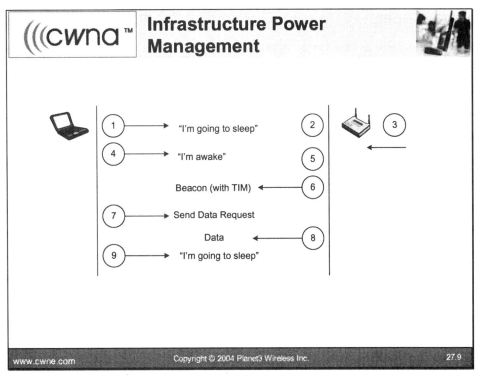

Sleeping stations only awake to monitor the beacon. If traffic is queued for them, they power up, notify the access point that they're awake, and receive their traffic. Then they go back to sleep. Stations wake themselves up. Access points cannot wake up a sleeping station since the sleeping station cannot hear the access point while asleep. Stations know when to wake up based on a chosen beacon interval.

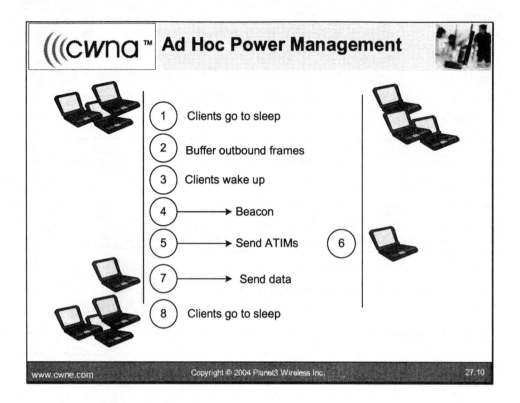

Ad Hoc power management is significantly different than infrastructure power management. Stations alternate sending beacons using a backoff algorithm. The beacon interval is the same as the ATIM interval. Stations send announcement traffic indication messages (ATIMs) to each other during a period of time called the ATIM window. During this window, a transmitting station notifies the receiving station that it needs to stay awake to receive data. The ATIM window immediately follows the beacon, and all ATIMs are ACKed by receivers. Stations send data during a data window then go back to sleep unless told to stay awake by another station with an ATIM.

Chapter 28
Security Mechanisms and Common Threats

 Objectives

Upon completion of this chapter you will be able to:

- Understand what security mechanisms the 802.11 standard specifies
- Recognize common WLAN security problems & attacks

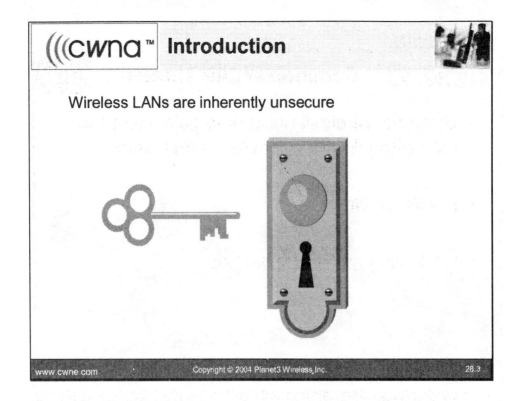

Wireless LANs are inherently unsecure and require a different perspective as compared to wired network security. Steps must be taken to secure wireless LANs in the same way steps are taken to secure wired networks. It is the responsibility of the CWNA to ensure proper wireless LAN security at their place of business or home.

The IEEE 802.11 standard by default does not provide for adequate security. It is currently difficult to physically identify or locate an intruder. Wireless LANs extend the edge of a network beyond the physical boundaries of a facility. Wireless devices present a security threat even for organizations that have no wireless implementations.

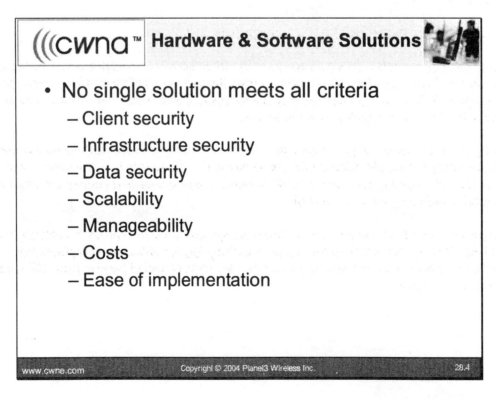

Common WLAN Threats

- Using the wireless network to gain access to both <u>wired</u> & <u>wireless</u> network resources

- Packet Sniffing

- DoS using an RF transmitter

Training is the key to security!

IT professionals need to understand how an intruder gains unauthorized access, understand the role of corporate security policy in securing a wireless LAN, and the importance of making sure that end users follow policy. IT professionals must also understand available categories of wireless LAN security solutions and how they work.

Common WLAN Threats

The most common threats to wireless LAN security include the following:

1. Using the wireless network to gain access to the wired network. A visitor with a laptop and wireless card uses your Internet connection from the parking lot of your office; a neighbor uses your home DSL connection through your access point; a hacker uses the wireless network to bypass the firewall and hack into wired servers.

2. Using wireless network to gain access to wireless resources. You are working on a document while waiting for a flight. Meanwhile, the hacker in the seat next to you is accessing your hard drive through your wireless card. Or, the person in the next cubicle is reading the email on your hard drive through your wireless card.

3. Capturing data off the wireless net. A corporate spy uses a wireless protocol analyzer to capture packets off of the WLAN from the company parking lot; the office snoop captures that "sensitive" email you sent regarding your boss; Wireless Denial of Service using RF transmitter or packet generator.

The Security Balance

- There is no "right" answer
  - How much will it cost?
  - Is there low-hanging fruit?
  - How difficult will implementation be?
  - What's the management overhead?
  - How useable is the solution?

There is no such thing as an unbreakable security mechanism; rather, there are only mechanisms that are increasingly difficult to break into. The question is not, "how am I going to keep hackers out?" In the long run, you cannot keep hackers out. The question is, "how am I going to make it so difficult for hackers to get in that it's not worth their time, and how am I going to catch them attempting to get in before they actually get in?" Every security mechanism that you add to the network makes it somewhat less convenient for legitimate users to get in.

A balance must exist between making the network hard for hackers to get into and making it easy for authorized users to get into. This balance is in a different place for every network, so there is no "right" answer to security. You *must* prevent your network from being on a hacker's list of "low-hanging fruit".

802.11 Security

- The 802.11 standard offers rudimentary security
  - Open System Authentication
  - Shared Key Authentication
  - Static WEP keys (for authentication & encryption)
  - MAC address filtering
  - SSIDs for network segmentation

The 802.11 standard offers rudimentary security, including:

- Open System Authentication
- Shared Key Authentication
- Static WEP keys (for encryption)
- MAC address filtering
- SSIDs for network segmentation
- Per-user authentication is not defined
- Rotating WEP key mechanisms are not defined
- VPN technology is not defined

Open System & Shared Key Authentication do not address:

- Per-user authentication
- Mutual authentication
- Dynamic WEP keys

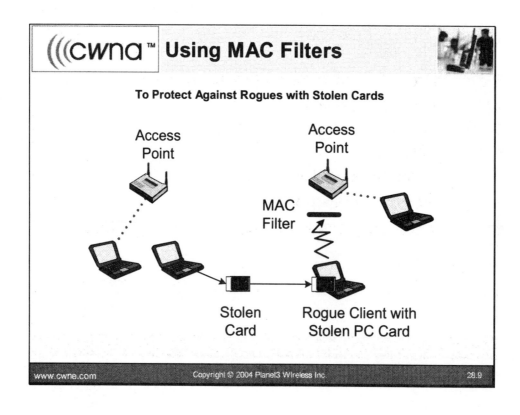

MAC filters permit traffic based on the MAC address of the sending station, and allow or deny traffic based on MAC address. MAC filters are easily circumvented by MAC address spoofing. Access lists generally must be manually configured, and do not allow for visiting users. Proprietary solutions for managing access lists exist, but are not widely implemented. Most access points have a maximum number of addresses allowed. The typical limit is around 256 stations. Large sites may need to exceed this limit. MAC filters do not address security in ad-hoc networks. MAC filters may be used effectively in bridging scenarios.

Service Set Identifier (SSID)

- Used as a network name

- Should <u>NOT</u> be used as a security mechanism

- SSIDs are broadcast in clear text in many frame types

The SSID is, or should be, used as a network name (such as "WORKGROUP" in a Windows network) in order to segment the network. Each access point and end station is programmed with an SSID. End stations can only access the access points with their same SSID. The SSID should *not* be used as a security mechanism. SSIDs are broadcast in clear text in beacons, probe requests, probe responses, association requests, and reassociation requests.

Examples of How Attacks Occur

- Broadcast monitoring (eavesdropping)
- Denial of Service
  - Jamming
  - Hijacking
  - Man-in-the-middle
- Rogue access points
- Accessing unsecured configuration interfaces

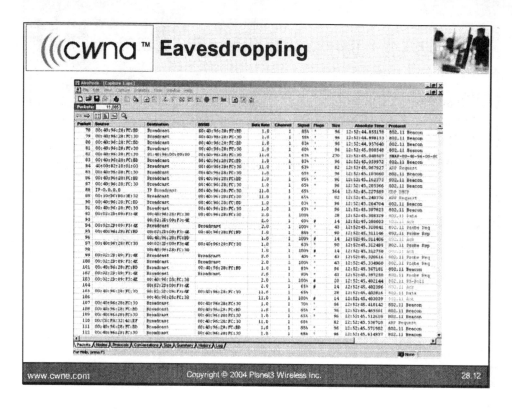

Casual & Malicious eavesdropping:

- Gathering login info using a protocol analyzer
- Monitoring sites visited
- Monitoring conversations (IM & Email)

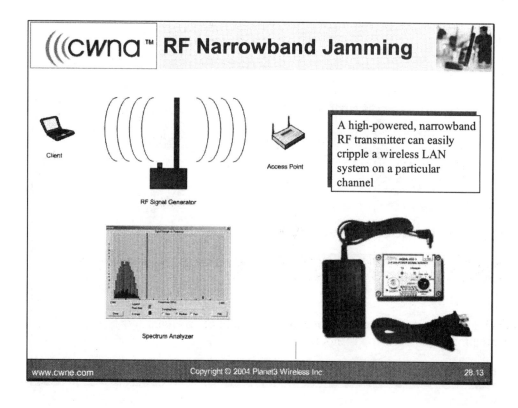

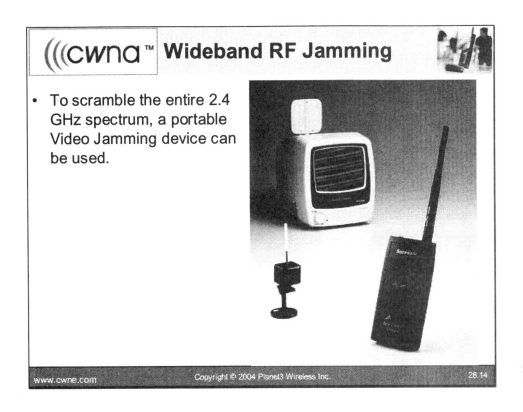

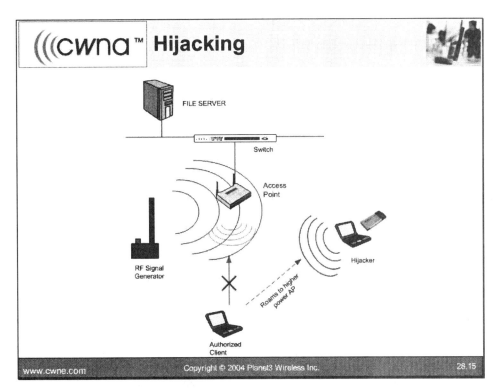

Wireless hijacking can be easily accomplished through use of access point software and DHCP server software loaded on a wireless laptop computer.

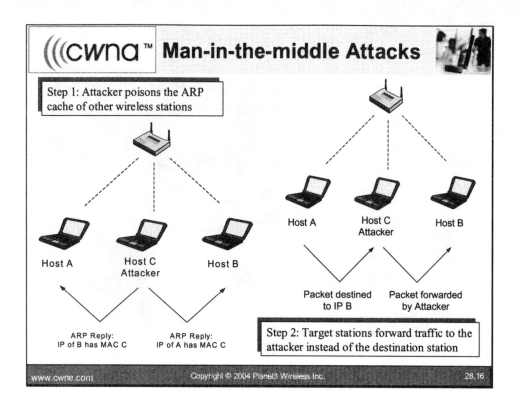

Chapter 29
802.1x/EAP, TKIP, WPA, 802.11i, and AAA

Objectives

Upon completion of this chapter you will be able to:

- Describe the use of dynamic WEP keys
- Explain 802.1x/EAP functionality and the types currently available in the market
- Describe Temporal Key Integrity Protocol (TKIP) and explain why it is used
- Describe Wi-Fi Protected Access (WPA) v1.0 & 2.0
- Explain how WPA relates to IEEE 802.11i
- Describe the Advanced Encryption Standard (AES)

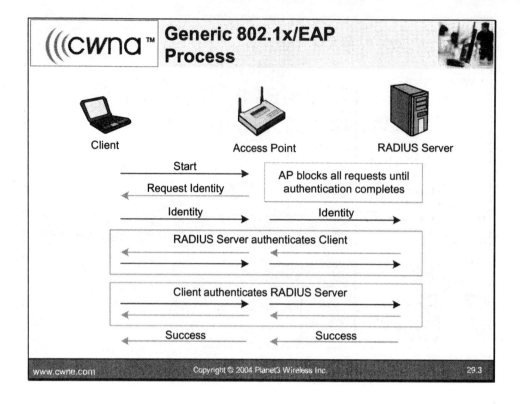

For added security, mechanisms exist for rotating WEP keys dynamically without user intervention. 802.1x/EAP solutions typically use an authentication server for this purpose. Some access points can perform this function internally. Key servers are usually RADIUS or TACACS+. User-based mutual authentication is always a good idea and is available with most EAP options.

802.1x Standard

802.1x is not an authentication method, even though people like to refer to it as one. 802.1x is a standard for port-based access control, and is primarily used in wired networks. 802.1x has been adapted for wireless networks, and is supported in RADIUS. 802.1x ties an authentication method (defined by the EAP type) to the use of the wireless media, and is outside the scope of the 802.11 standard. 802.1x is supported by many enterprise wireless hardware and software products.

802.1x prevents unauthenticated users from establishing a Layer 2 connection. It suggests but does not require encryption, and is supported natively by Windows XP. 802.1x utilizes three parts:

1. Supplicant (client)
2. Authenticator (access point)
3. Authentication Server (RADIUS server)

Access points supporting 802.1x act as a pass-through between the client and authentication server.

Authentication, Authorization, & Accounting (AAA)

RADIUS & TACACS+ usually perform AAA functions. AAA Servers are used with wireless networks for user authentication and other functions. The AAA Server may rotate and distribute dynamic WEP keys. AAA Servers are used with both 802.1x/EAP and VPN solutions.

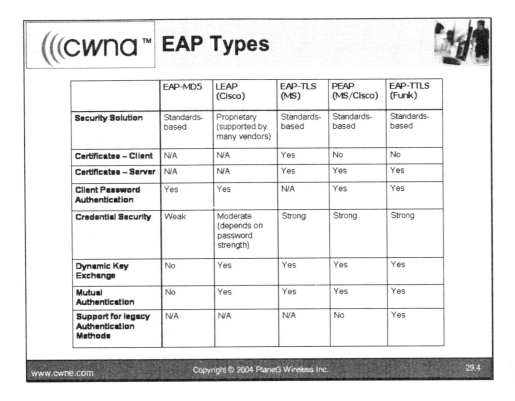

EAP, Extensible Authentication Protocol, is a Layer-2 authentication protocol used over Ethernet as a flexible replacement for PAP and CHAP under PPP. Windows XP supports most versions of EAP starting with service pack 1. Service pack 2 with patch 815485 supports WPA (Wi-Fi Protected Access).

There are many types of EAP – each having its own advantages and disadvantages, which are explained in the following pages.

- EAP-MD5
- EAP-Cisco Wireless (LEAP)
- EAP-TLS
- EAP-TTLS
- EAP-PEAP

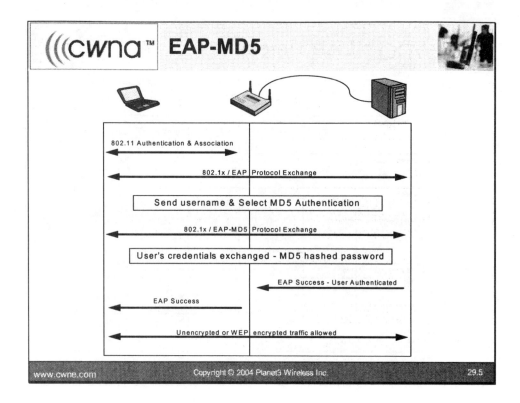

EAP-MD5 is standards based, and is called for in RFC2284 for 802.1x. However, EAP-MD5 has three weaknesses:

1. One-way authentication
2. Challenge passwords
3. No per-session WEP keys

Also, EAP-MD5 is based on CHAP authentication, and is rarely used due to weak security.

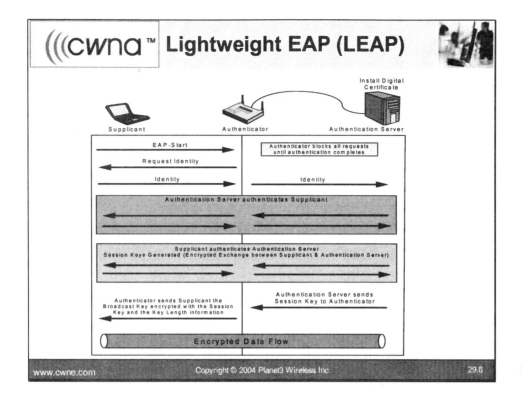

LEAP offers a moderate authentication security level based on password strength. LEAP is Cisco proprietary, but CCX is now available to the market for client and AAA development. LEAP was the first EAP type to gain acceptance in the market, and is now supported by many vendors including Apple and Intel Centrino. LEAP is based on a modified version of MS-CHAPv1 and uses only password authentication, and is simple to implement and manage. The username is sent across the wireless medium in clear text.

Joshua Wright (http://home.jwu.edu/jwright/) has developed an offline dictionary attack against LEAP called "ASLEAP". ASLEAP can crack LEAP passwords in seconds using the existing version of LEAP. Cisco is currently working on a fix for this weakness, but suggests use of EAP-TTLS, PEAP, and/or strong passwords as an interim solution.

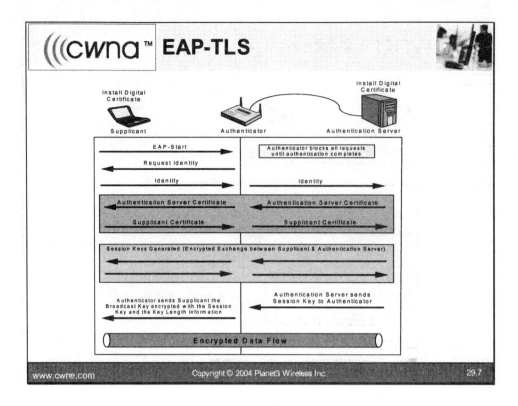

EAP-TLS offers strong authentication security. EAP-TLS is based on SSLv3, uses client-side and server-side certificates, and is used as a foundation for newer types of EAP solutions such as PEAP and EAP-TTLS. EAP-TLS is best implemented when there is an existing Public Key Infrastructure (PKI) because of its high management overhead.

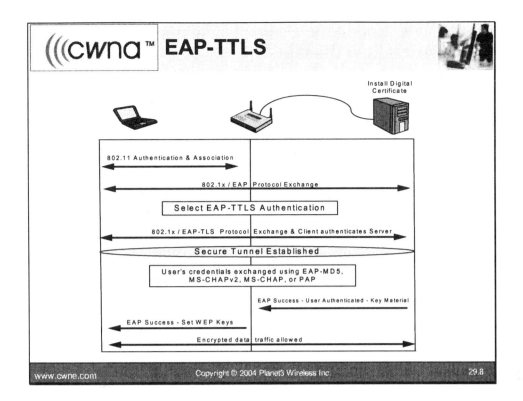

EAP-TTLS offers a high level of authentication security. EAP-TTLS was developed by Funk Software and Certicom, and is supported in Funk's Odyssey software. EAP-TTLS builds an encrypted TLS tunnel between the supplicant and authenticator before user credentials are passed to the authentication server. EAP-TTLS competes against PEAP using server-side certificates and client-side passwords. EAP-TTLS can use a range of authentication methods inside the encrypted tunnel, and offers low management overhead.

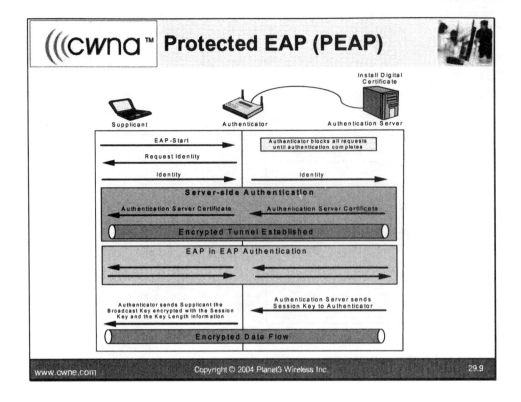

PEAP, which was developed by Cisco, Microsoft, and RSA, offers a high level of authentication security. PEAP was supported starting with:

- Windows XP service pack 1
- Cisco ACU version 5.05
- Cisco ACS Server 3.1
- Meetinghouse AEGIS client and server
- Funk Steel-Belted RADIUS v4.04

PEAP competes against EAP-TTLS using EAP within EAP. PEAP builds an encrypted TLS tunnel between the supplicant and authenticator before user credentials are passed to the authentication server, and has low management overhead.

Microsoft supports:

- PEAP-EAP-TLS (server certificates / client certificates or smartcards)
- PEAP-EAP-MS-CHAPv2 (server certificates / client passwords)

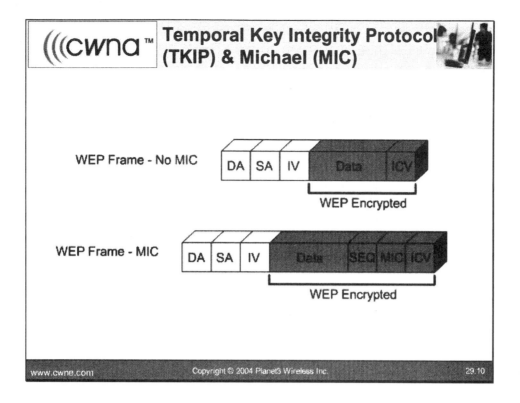

Temporal Key Integrity Protocol (TKIP) & Michael (MIC) were developed by Cisco as a solution for the known weaknesses in WEP. TKIP was adopted by the Wi-Fi Alliance as an immediate solution until 802.11i is ratified. TKIP starts with a 128-bit temporal key, adds the MAC address and a 48-bit IV (Initialization Vector, extended from the original 24 bits) to create the encryption key (key mixing). A new Message Integrity Check (MIC) called "Michael" is added. The temporal key is changed automatically after every 10,000 packets. TKIP uses RC4 for encryption just like WEP, and is easily implemented via a firmware update on existing hardware devices.

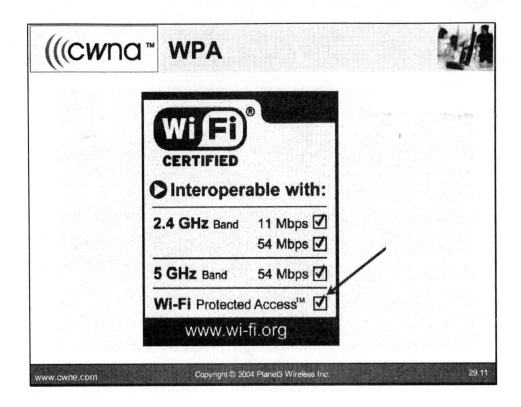

WPA is designed to run on existing hardware as a security upgrade firmware patch. The goals for WPA are:

1. Strong data encryption through TKIP
2. User authentication through 802.1x/EAP with mutual authentication

Mixed mode WPA (simultaneous WEP and WPA clients in one BSS/ESS) is discouraged by the Wi-Fi Alliance because this mode of operation is inherently unsecured, but many vendors support it. WPA Certification testing includes a negative test to ensure that the device under test does not support mixed mode in its default configuration.

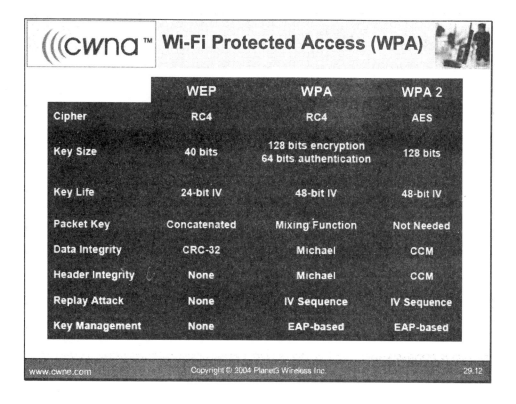

WPA v1.0 is a subset of the forthcoming IEEE 802.11i standard. WPA has two modes:

1. WPA 1.0 defines a "Pre-Shared Key Mode" where TKIP and a pre-shared key are used for simplicity of deployment
2. WPA 1.0 defines 802.1x/EAP with use of TKIP

WPA v2.0 is forthcoming and will be fully 802.11i compliant. The Wi-Fi Alliance will require mandatory compliance with WPA in order to maintain Wi-Fi status by Q4 - 2003. WPA v1.0 does *not* include the following 802.11i items:

- Secure IBSS
- Secure fast handoff
- Secure de-authentication and disassociation
- Advanced Encryption Standard

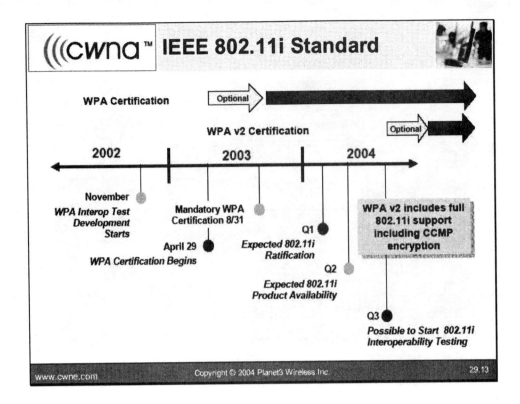

The 802.11i committee is specifying:

- Use of TKIP
- Use of 802.1x / EAP
- User-based authentication
- Mutual authentication
- Use of AES
- Due for release Q2 – 2004

AES is a replacement for the RC4 stream cipher used in WEP, and uses the Rijndael Algorithm. AES uses key Lengths of 128, 192, 256, and is considered *uncrackable*. AES is specified in forthcoming 802.11i standard, will be part of WPA 2.0, and will require new hardware (client, AP, Bridge, etc.) with more power, rather than just a firmware upgrade.

Chapter 30
Baseline Practices, Common Security Solutions, & Regulations

Objectives

Upon completion of this chapter you will be able to:

- Explain baseline security practices
- Common VPN security solutions
- Industry regulations

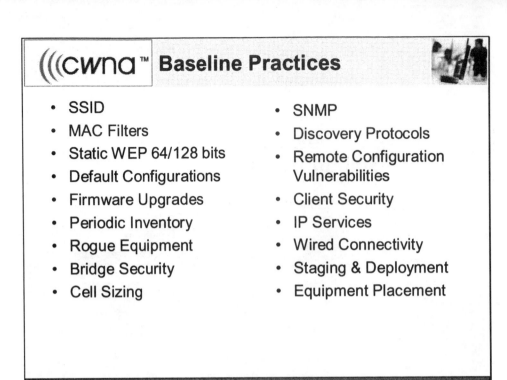

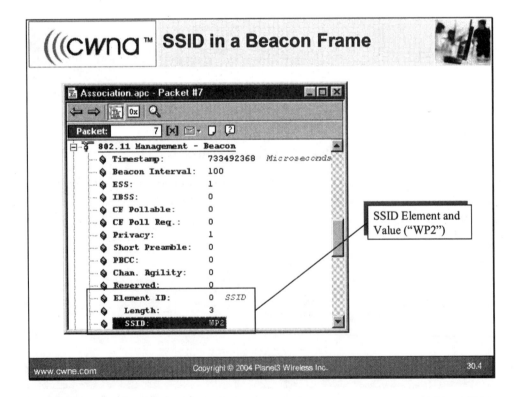

Do not broadcast SSIDs in beacons, probe response frames, or others where configurable. Do not allow clients using a broadcast ("any") SSID to associate. Do not use manufacturer defaults for SSIDs. Do not use SSIDs that indicate corporate info. Add a null (blank) character at the end of an SSID whenever possible. Packet analyzers have to be set to specifically look for a null (blank) character (hex mode), and NetStumbler will not see it.

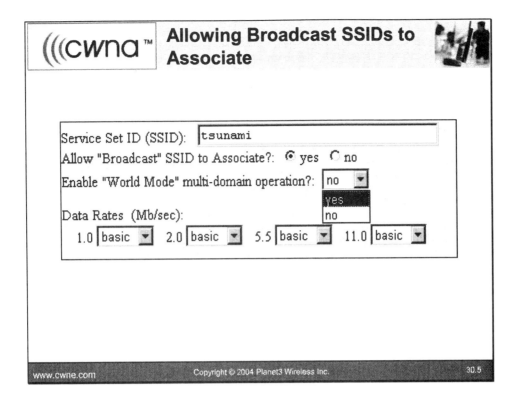

The illustration is an example of a configuration dialog from an access point. Notice that it has been set to allow the broadcast SSID to associate.

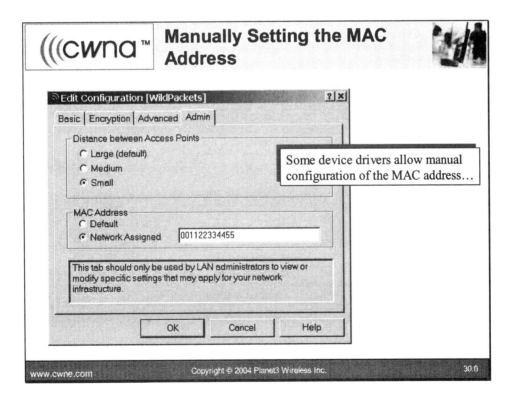

Active MAC addresses can be easily located in a packet capture. MAC addresses are easily spoofed using common freeware utilities.

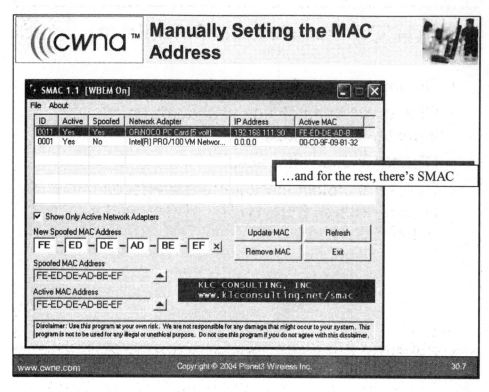

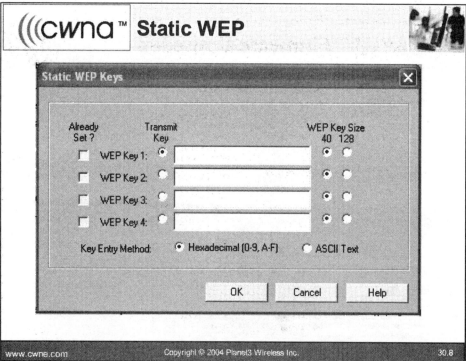

When using static WEP, use strong keys unrelated to:

- Organization's name
- SSID
- Organization's address or phone number
- Access point's or bridge's model number
- Manufacturer defaults

Use WEP for basic, economical means of security in a SOHO environment. Use the highest level of WEP encryption supported by your hardware.

Manufacturer Defaults

- Do not use manufacturers' default settings
- Change usernames and passwords on infrastructure devices
- Disable unused interfaces
- Password protect management interfaces such as Telnet, SSH, HTTP, HTTPS, and SNMP

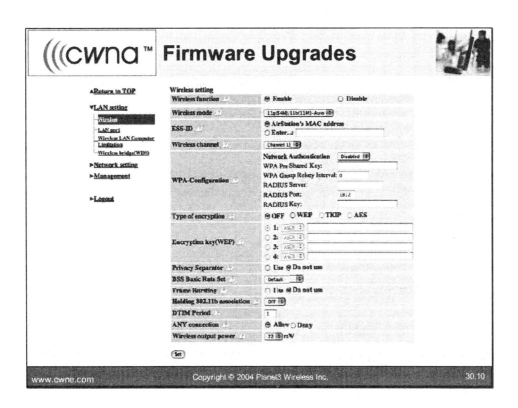

Upgrade firmware on access points, bridges, clients, workgroup bridges, and Enterprise Wireless Gateways. Access latest security features support for WPA 1.0, TKIP, AES, Kerberos, 802.1x / EAP (TLS, TTLS, LEAP, PEAP).

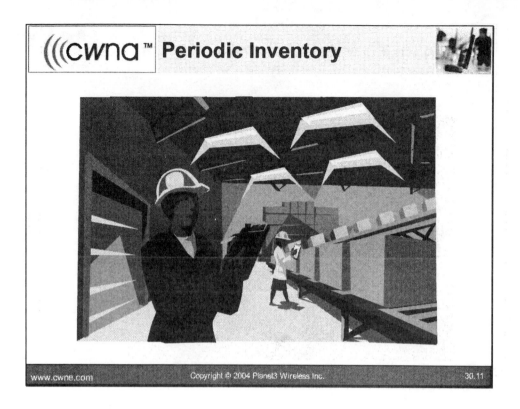

Employees "lose" their cards which can compromise a network built on static WEP as the only security measure. Some organizations may be too large to inventory, necessitating an automated inventory process. Some organizations may not be able to implement an inventory procedure at all.

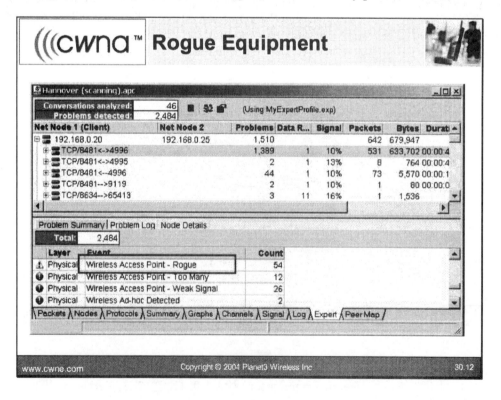

Treat rogue equipment as a serious breach of network security. Eliminate rogue wireless equipment through network administrator training, end user training, intrusion detection systems, and regular and continuous searches for all available wireless technologies. After finding a rogue, do a complete sweep of the network looking for setting changes. Implement a wireless management system that can detect improperly configured infrastructure devices.

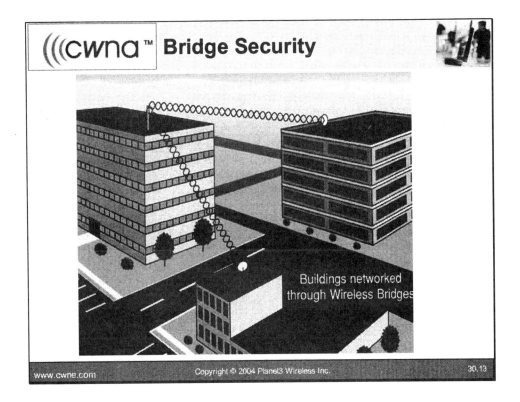

Understand before implementation that bridge links can span miles so an attacker will never be seen. Disable a bridge's access point features when not in use. Do not allow clear text transmissions. Traffic passed between wireless bridges in the clear can be captured by a packet analyzer from miles away from the hardware. Rogue bridges may be placed onto the network at a range of several miles. A good security solution MUST be determined and tested prior to implementing a wireless bridge link. Authentication and encryption over the bridge link are essential.

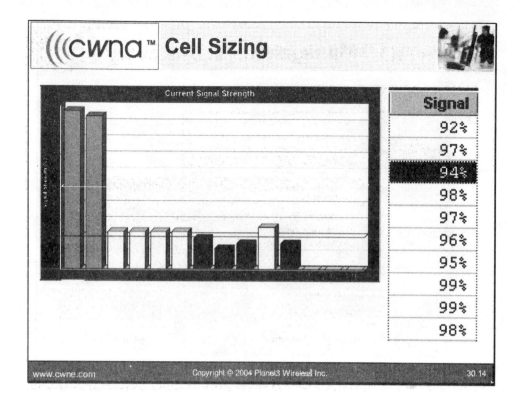

Limit your wireless LAN's footprint by limiting the output power to only what is needed. Use antennas wisely to avoid signal spillover to outside the facility. Perform footprint analysis with omni and directional antennas.

Examine the signal strength as you move between various locations. Assess the feasibility of accessing your wireless network from various locations. If the signal bleeds out into the parking lot, then a company policy preventing access to the parking lot may increase the security of your wireless network.

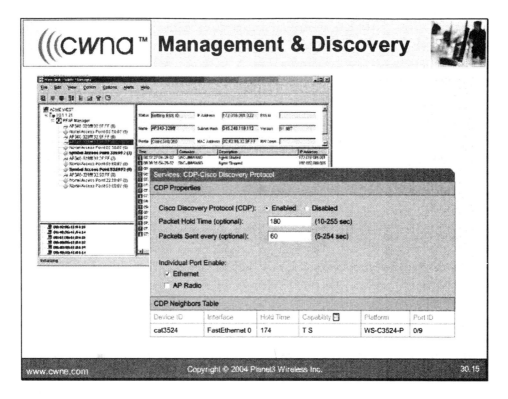

Steps that should be taken regarding use of SNMP include:

- Turn off SNMP service if it is not being used
- Use strong strings
- Do not use default strings
- Do not use anything related to the SSID, WEP key, or organization
- Do not pass SNMP strings in the clear over a wireless segment
- Use SNMPv3 with encrypted community strings whenever possible
- Disable Layer-2 discovery protocols (like CDP) when not in use
- If possible, assure access points and bridges cannot be configured over the wireless segment
- Never log into access points and bridges over the wireless segment unless the wireless link is encrypted

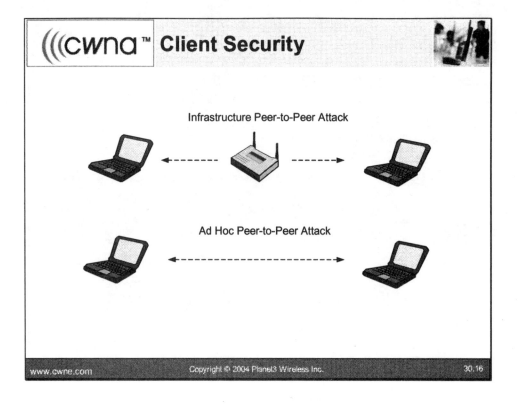

Because peer-to-peer attacks on wireless clients are extremely simple for an experienced hacker to conduct, always limit sensitive data on wireless clients, prohibit and prevent use of corporate computers on public access wireless networks without protection, and use personal firewall software or IPSec policies on workstations. Also, eliminate file and folder sharing on workstations, and ensure that users getting into your network over an unsecured connection, like a VPN, do not have unsecured wireless at their remote location (usually home or remote office).

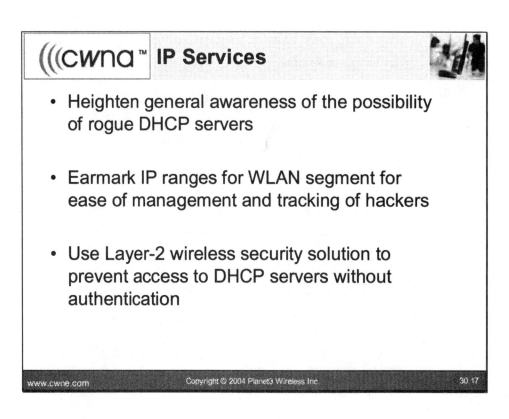

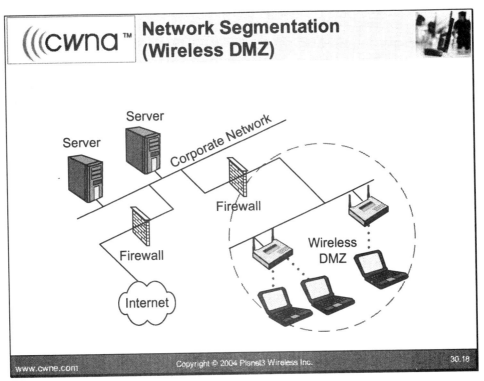

Use switches instead of hubs for wired connectivity. Switches offer 802.1q VLAN tagging, segmented network design, and full-duplex connections.

The wireless segment can also be protected using various network segmentation devices, such as firewalls, L3 switches, wireless LAN switches (using VLANs), Enterprise Wireless Gateways (EWGs), Enterprise Encryption Gateways (EEGs), and VPN Concentrators/Gateways.

Configure and test infrastructure devices in an isolated environment prior to deployment. Use an approved staging checklist to ensure that all security and administrative policies are met consistently.

Copyright 2004 Planet3 Wireless, Inc.

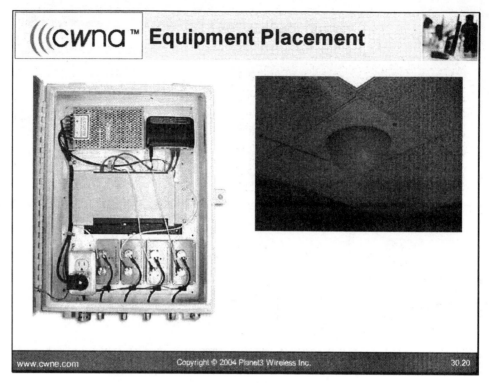

Secure wireless infrastructure hardware to avoid physical theft, replacement, or tampering. Secure console connections to wireless LAN infrastructure devices. Lockable ceiling-mounted enclosures or lockable NEMA enclosures work well.

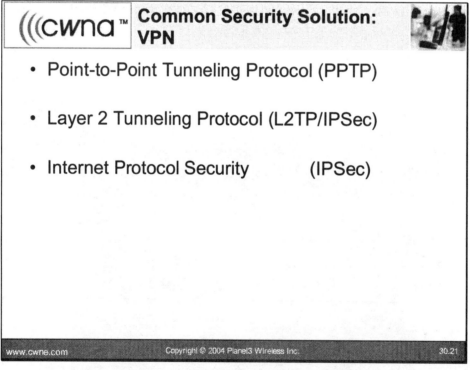

VPNs are a common security solution, including Point-to-Point Tunneling Protocol (PPTP) and Internet Protocol Security (IPSec).

Consider Layer 2 Tunneling Protocol (L2TP) with IPSec, and Secure Shell v2 (SSH2). VPNs are understood by most security professionals because VPNs are a mature solution that is already deployed and in place. VPNs provide for user-based authentication & strong encryption, and are commonly implemented in Enterprise Wireless Gateways (EWGs) and firewalls.

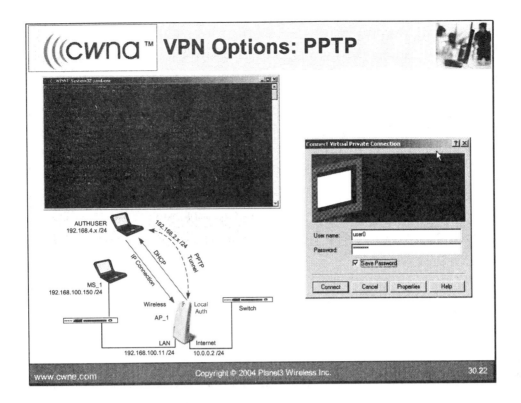

Developed by Microsoft and supported by many other vendors, PPTP is a VPN protocol that is simple to configure in a client/server model. PPTP is fairly secure with MPPE-128 encryption. PPTP is supported in most Windows versions, and also available in freeware servers (Linux POPTOP). PPTP is available in commercial network appliances such as SnapGear and Symantec, and is supported natively in Windows 2000 Server.

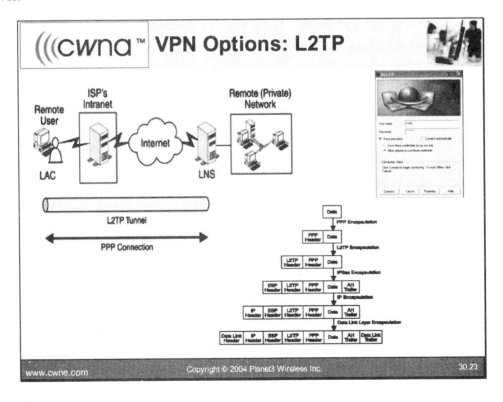

L2TP was co-developed by Cisco and Microsoft. L2TP is a hybrid of Layer2 Forwarding (L2F) and PPTP to have the best features of both protocols. L2TP is fairly simple to configure, but not as widely supported as PPTP, though it supports IPSec security.

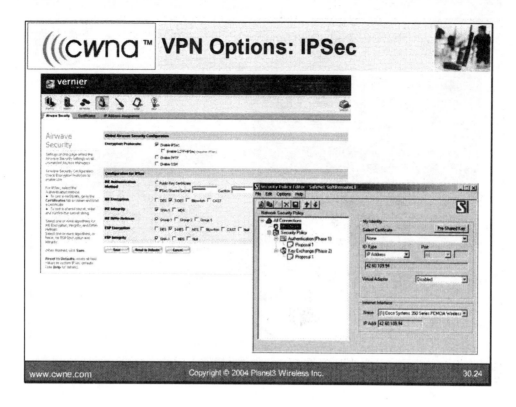

IPSec offers very strong security through 3DES & AES encryption and through a configured policy. IPSec is moderately difficult to configure, and is supported by many vendors. IPSec is the industry standard for high-end security.

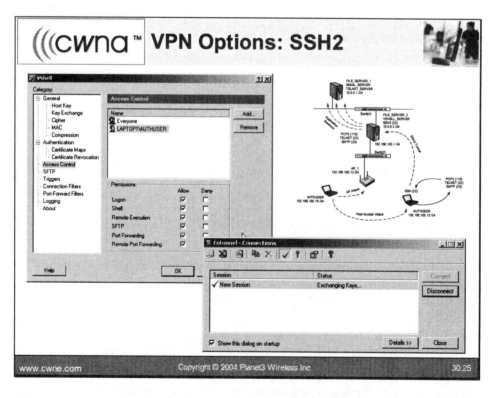

SSH2 Tunneling requires client and server software, and is simple to configure. SSH2 creates moderate overhead, but is very secure. SSH2 uses Public Key / Private Key encryption scheme, and is low cost. SSH2 Port redirection can be handled "locally" on the client computer. Redirected ports must be preconfigured in client software, and can be handled "remotely" on the server computer.

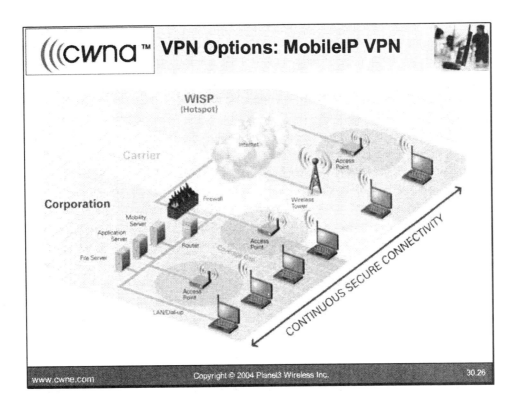

Mobile IP VPN solutions are built on RFC 2002, which is still in its infancy. Mobile IP is a Client/Server software solution, in which different vendors are often interoperable. The speed of the solution depends on the implementation. Mobile IP supports RADIUS, and offers flexibility in roaming across Layer 3 boundaries.

There are multiple parts to a solution (home agent, foreign agent, client software). Most solutions support IPSec encryption, and some Mobile IP solutions may provide application persistence.

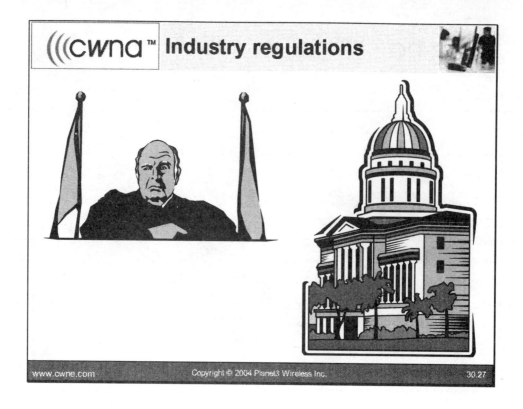

With the exception of the federal government, regulatory standards are not linked to specific technologies. When implementing a wireless LAN, be aware of laws and regulations that are specific to certain industries.

HIPAA – (Health Insurance Portability and Accountability Act) – Businesses in the health sector must protect PHI (protected health information) and implement procedures and policies to safeguard patient information in all formats including electronic.

GLBA – (Graham-Leach-Bailey Act) – Banks and other financial institutions must guard the security and confidentiality of a customer's "personally identifiable" financial information.

FIPS – (Federal Information Processing Standard) – Federal agencies must abide by specific technology security requirements. For example, FIPS-140-2 calls for use of AES encryption.

Sarbanes-Oxley – Public companies must establish and maintain adequate internal controls over their financial reporting systems to prevent fraud.

Chapter 31
Site Surveys: Administrative Perspective

Objectives

Upon completion of this chapter you will be able to:

- Understand and explain the need for a site survey
- Determine the business justification for a wireless LAN
- Understand the customer's network topology
- Document site survey results on the appropriate forms
- Define and create an RF site survey report

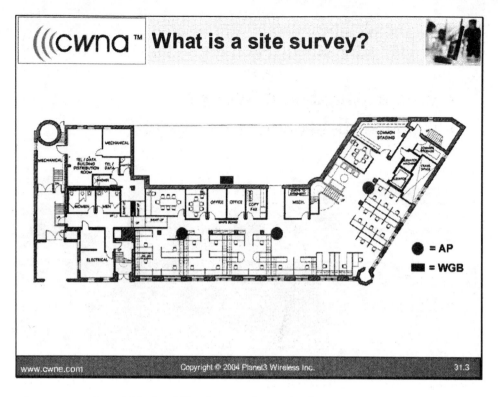

A site survey is a process by which the following are determined:

- Feasibility of the desired RF coverage in a given area
- Appropriate wireless LAN hardware placement
- Wired connectivity limitations

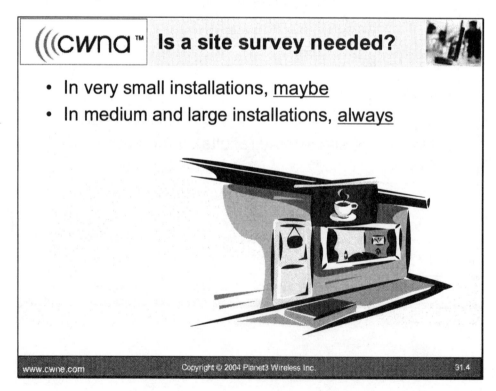

When is a site survey required? A site survey might be done in very small installations, such as small office / home office, coffee shop, or fast food restaurant. A site survey should always be done in medium and large installations, such as site-to-site bridge links, warehouse, hospital, or multi-tenant buildings.

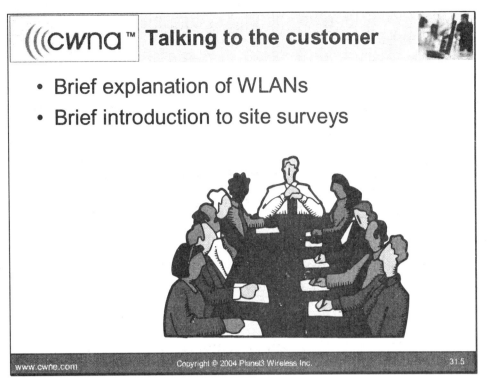

It is often necessary to explain to the customer the differences between wired and wireless LANs. Many customers will need a brief introduction to site surveys in order to understand the need for the service.

Categories of questions to ask include:

- Purpose of the wireless LAN
- Business requirements
- Security requirements
- Available resources
- Existing networks

Purpose of the Wireless LAN

- Why is the organization considering a wireless LAN?
- What will the wireless LAN be used for?

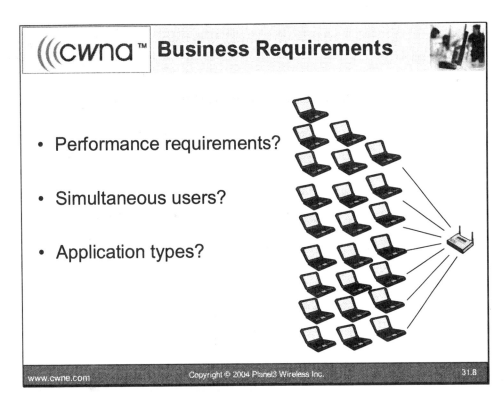

In gathering the business requirements, the following questions are recommended:

- What performance requirements are expected?
- How many users will be on the wireless LAN at the same time?
- Do all users require connectivity or only specific users or groups?
- What type of applications will be used?
- Telnet, SSH, Citrix, HTTP, POP3, CAD/CAM, FTP, SCP, TFTP, File Server access
- Are there bottlenecks like an Internet T1? If the wireless LAN is primarily used for Internet access, it's hard to justify anything faster than 802.11b (11 Mbps).
- Is fault tolerance needed? Wireless LAN switching can provide automatic power adjustments to active access points when an access point fails. Multiple access points may be placed in a particular area in load-balancing or hot-standby configuration.

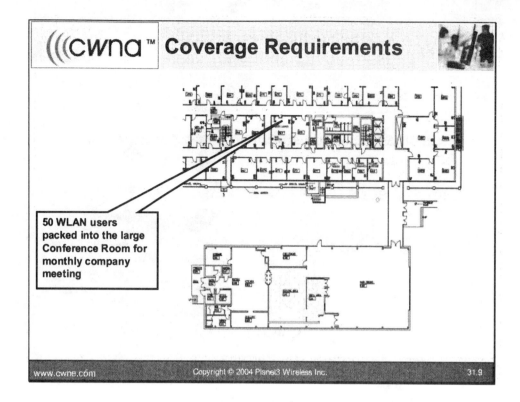

Are there specific locations or is general coverage needed everywhere? Are there any special circumstances? In conference rooms and hot spots, users are not mobile and may desire higher throughput.

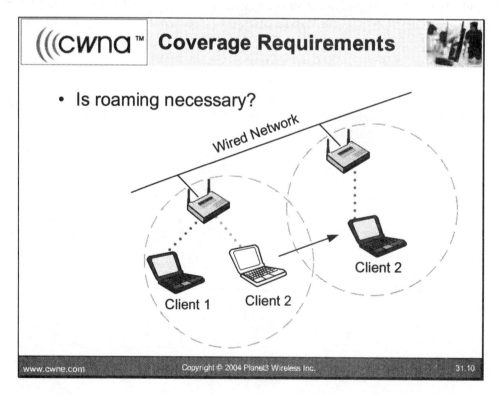

Installing a wireless network just to have one is not a cost-effective idea. Giving network users mobility is an appropriate reason to implement a wireless LAN.

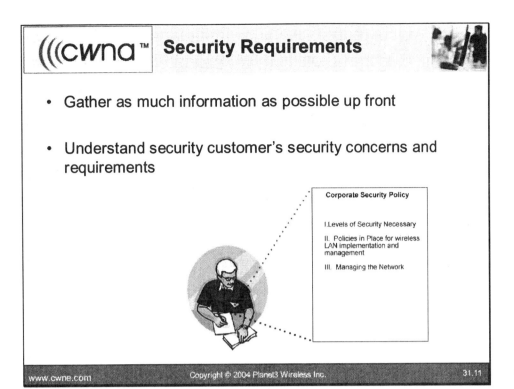

It is wise to gather as much information up front as possible while you have the opportunity. Security concerns will be addressed with the customer as a separate item from the site survey. Remember that the site survey covers hardware placement, RF coverage, and interference discovery. While wireless security is a primary concern, it is only related to the site survey in that an equipment vendor should be chosen that meets both the site survey requirements *and* the security policy requirements of an organization.

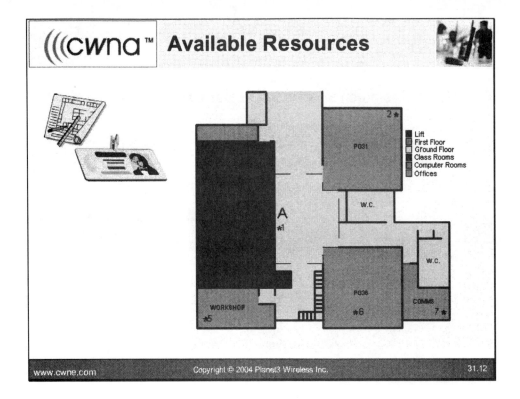

Some new wireless LAN switch vendors require blueprints/floor plans in order to do the survey because the site survey process is mostly automated.

- Are blueprints available?
- Are any previous site survey reports available?
- Is a badge required?
- Are the wiring closets accessible?

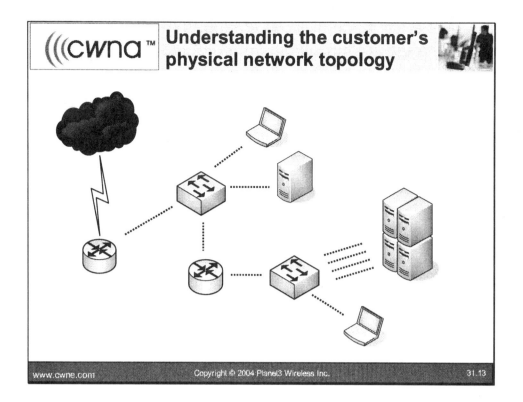

Understanding the customer's network allows the site surveyor to make wise decisions about where an access point can and cannot be installed. Wiring closet locations, Layer-3 boundaries, and DMZs are a few items to consider when determining appropriate physical and logical locations of access points. Obtaining a network topology map from the customer will save the site surveyor significant time and trouble. Depending on corporate security policy, some customers may not allow the site surveyor access to the network topology map.

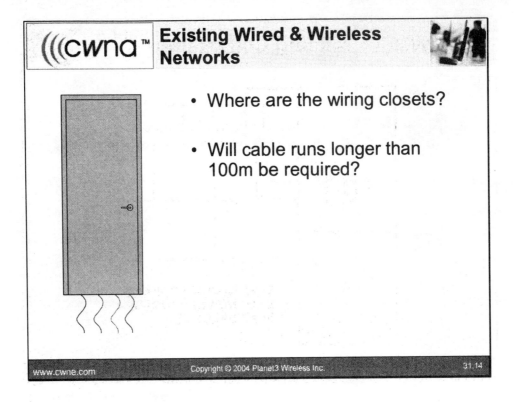

- Where are the wiring closets?
- Are hubs or switches in use at the network edge?
- Is there an existing wireless LAN?
- Does it function?
- What frequencies and technologies does it use?
- Is there PoE capable switches or multi-port injectors at the network edge?
- What voltages are supported?
- If cabling runs are greater than 100 meters, are fiber connections available in the edge switches?
- What protocols are in use on the network?

Remember that fiber-connected access points must be AC powered. This information might be needed for choosing equipment and configuring protocol filters.

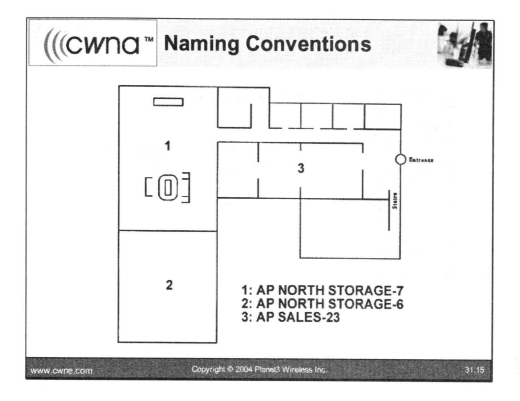

Has an access point/bridge naming convention been established? Network management is much easier when using names that indicate location or departmental use.

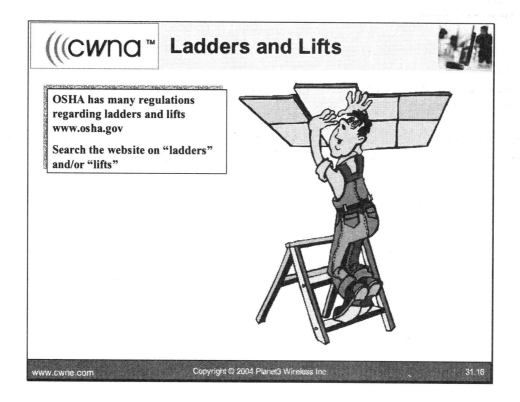

Will manlifts or ladders be required during the survey? Who will provide them? Is a license or certification required to drive the lift?

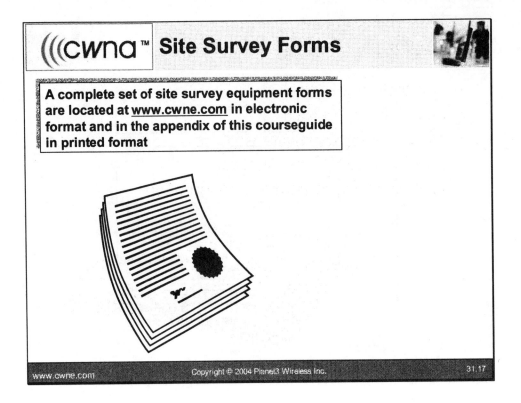

Forms should have specific details on where equipment goes and how it is to be mounted and powered, including:

- Antenna orientation (especially for directional antennas!)
- Access point orientation and mounting hardware
- Access point power sources (AC, PoE)

Do not use temporary objects as markers in your photos and documents because they could be moved before equipment gets installed. The necessary forms for documenting a site survey properly include an individual form for EACH access point, wireless bridge, and wireless workgroup bridge.

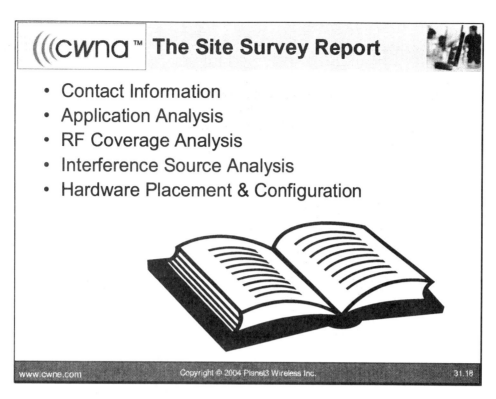

Contact Information

The IT manager interview form is used for collecting contact information of the customer before the site survey is performed. The site survey report should include complete contact information for the site surveying firm, the customer, and any third parties involved in the process. The main content of the site survey report should include:

Application analysis

An application analysis is almost always a customized testing process because all network environments are unique, and applications differ. Without adequate throughput, applications will not function properly, even if coverage is perfect.

Interference source analysis

- Where is the RF interference coming from?
- Can it be eliminated or blocked?
- Can we work around it?

RF coverage analysis

- Is the coverage where it needs to be to meet business goals?
- Where are the gaps, and can we work around them?

Hardware placement, configuration, and power information

- Where are the access points, bridges, & workgroup bridges located?
- How will access points, bridges, and workgroup bridges be powered?

** These topics are typically part of the site survey report, but other topics may be included as needed. Keep in mind that a *detailed* site survey report is a good report. You may be the surveyor, but not the installer.

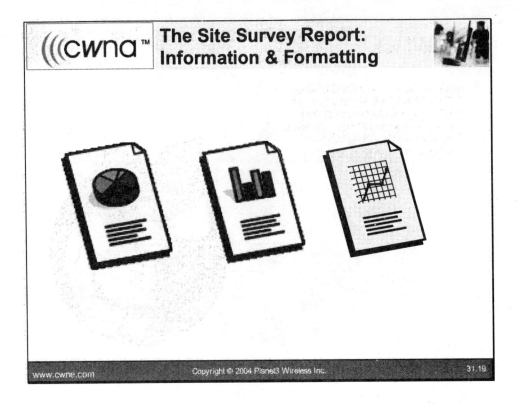

All information gathered during the site survey should be consolidated and detailed in the site survey report. Every consulting company has a different format for the site survey report, and this format is generally protected intellectual property. The document structure is much like any professional consulting report except that detailed photos are included. The report should include specific information such as equipment settings and survey tools and methods utilized.

The final site survey report usually has at least 1-2 pages for each access point, wireless bridge, or workgroup bridge, with the following details:

- Location, IP address, antenna type, mounting
- Cabling, 802.11 technologies, output power
- Power and data run type, digital photos
- Detailed graphical representation of the facility is mandatory
- Marked-up photo copies of printed floor plans
- Coverage areas and gaps
- Required coverage areas (and those areas that are not required)
- Automated graphical reports generated by site surveying software
- Often created by importing AutoCAD floor plans

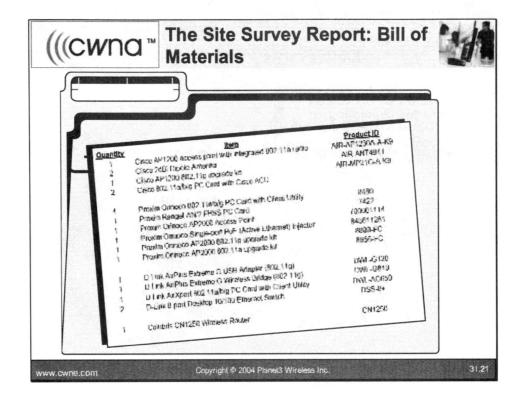

The site survey report may include a bill of materials, provided the equipment vendor is chosen by the customer before the site survey begins. Part numbers, descriptions, and costs of all components should be included. The actual installation should still be considered a separate function from the site survey.

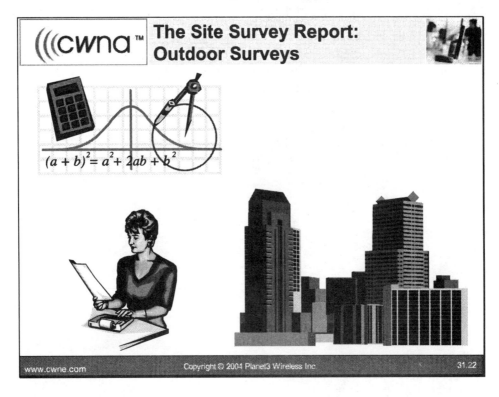

In an outdoor site survey report, include at least the following:

- Terrain maps and bridge locations
- Mounting & tower specifications
- System Operating Margin calculations
- Antenna, amplifier, cabling, and grounding specifications

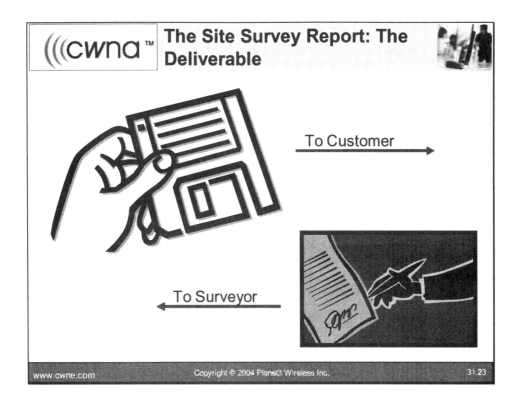

After the site survey is finished, it is common for the consulting firm to take 1-2 weeks to submit the report to the customer. Verify that the site survey report is dated. Verify that the customer signs for the site survey report – approving that the report meets their requirements upon delivery. The report is often printed and mailed to the customer, and often emailed and/or ftp'd to the customer in .zip format. A complete report will include .doc files, .pdf files, image files, and AutoCAD files.

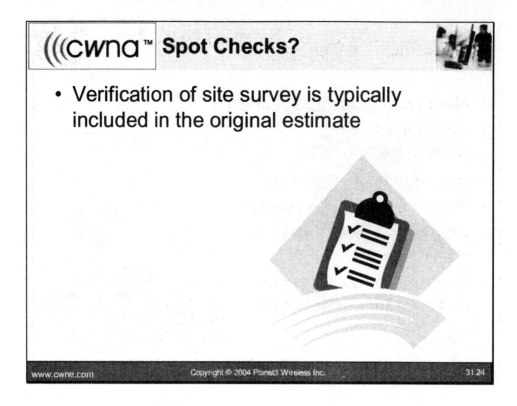

Site survey companies rarely want to be held accountable for their survey's accuracy. After a survey, the installation should prove or disprove the accuracy of the site survey. Someone – usually the wireless LAN owner – ends up doing the spot check to verify proper coverage and seamless roaming (if required).

What does spot checking include?

- Locate and document "dead zones" and areas where signal quality degrades significantly. Match these with the site survey.
- Watch for drop-outs between cells and how it affects prominent applications in use
- Watch for co-channel and adjacent channel interference
- Watch for excessive channel bleed-over
- Verify cell sizes and match against site survey

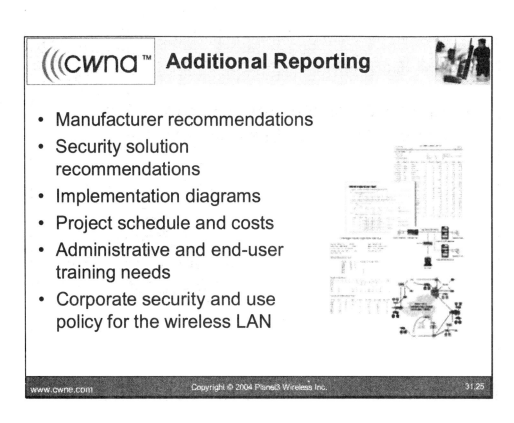

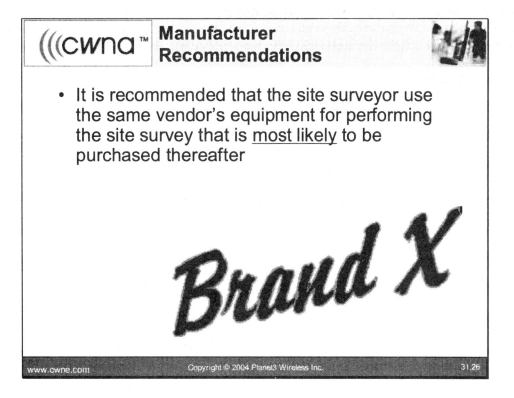

The 802.11 standards do not cover communication between access points across the distribution system. It is recommended to use the same vendor for all access points when seamless roaming is required.

It has been common practice in the industry for years to oversell wireless LAN equipment. When 30 access points are needed for adequate coverage and throughput, 60 access points are installed by a consultant. Too many access points are not good because the access points interfere with one another. Educated consumers are the best defense against this tactic.

Chapter 32
Site Surveys:
Technical Perspective

Objectives

Upon completion of this chapter you will be able to:

- Understand what hardware and software is typically needed to perform a site survey (site survey kit)
- Locate and avoid RF interference
- Determine RF coverage contours
- Locate appropriate placement for hardware installation
- Understand specific issues associated with surveying in various vertical markets and how to deal with them
- Understand FCC/FAA regulations concerning towers

It is often necessary to have multiple survey kits – one from each vendor that will be installed by your company. Survey kits can be expensive and have many parts so travel cases are usually required both for shipping and protection. Survey kits should consist of wireless LAN equipment covering 802.11a/b/g technologies (depending on vendor) because facilities vary greatly.

Access Points

- Carry at least two in case one malfunctions or gets broken
- Should support 802.11a/b/g technologies according to what types of connectivity the customer has requested

Battery Pack

- Cables made to attach to various vendors' access points
- Powerful enough to run the access point for a few hours
- Quickly rechargeable
- Have a charger on hand
- Have at least 2 batteries

Antennas

- Two of every kind you may have to use if you plan to use antenna diversity
- Use antennas made to work with the access point you're using for the site survey
- Survey with the antenna you plan to install later if possible
- If one antenna does not give appropriate coverage, do not guess what another antenna would do – test it

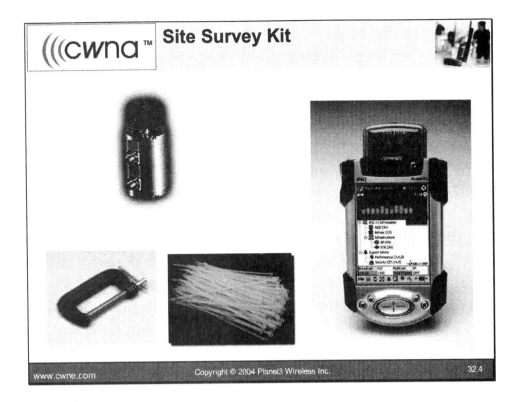

Cables and connectors to connect antennas to the test access point are necessary. Keep in mind that longer cables, pigtails, lightning arrestors, and connectors reduce signal power at the antenna. Use the same cables, connectors, splitters, and lightning arrestors during the survey as will be used during installation or use an attenuator to simulate the loss.

Client Device

- Whether a handheld or laptop PC is used, make sure to have extra charged batteries and a battery charger
- Lightweight/small units are preferred since surveys take hours and require significant walking
- Appropriate hardware/software
- Packet analyzer
- Site monitoring software (client utilities)
- Specialized devices

It is a good idea to survey with the same type of client device that will be used, being sure to record data rates, channels, packet sizes, etc.

Temporary mounting equipment

- Double-sided tape, Velcro, or electrical tape
- Brackets or beam clamps
- Zip ties, strong/thin wire, and paper clips

Be careful not to leave marks from temporarily mounting test access points.

Markers

- Bright-colored tape, signs, or stickers for marking access point and antenna placement in the facility
- Should be resistant to grease, dirt, dust, and water
- Must be easily removed without leaving a mark
- Digital pictures should include the markers for the site survey report

Digital Camera

- Used for documenting marked hardware placement locations
- Used for documenting unusual situations
- Try to include as much of the surroundings as possible when photographing an area. This aids in giving the installer the right perspective.

Distance Wheel

- Distance wheels are handy for determining Cat5 run distances
- Marked rope is good for noting vertical distances from the ground for hardware placement
- Laser range finders are often used outdoors

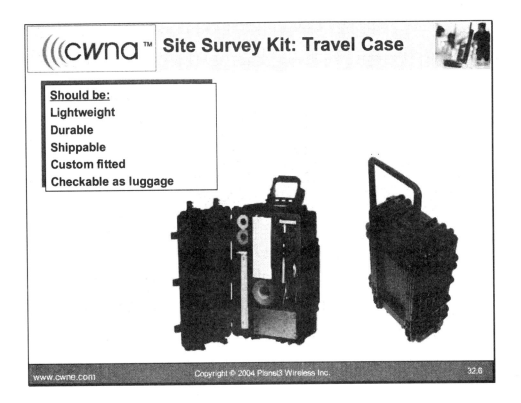

Custom travel cases will ease the burden of shipping/carrying so much gear and will protect the kit from damage. These cases should be checkable as luggage on airlines.

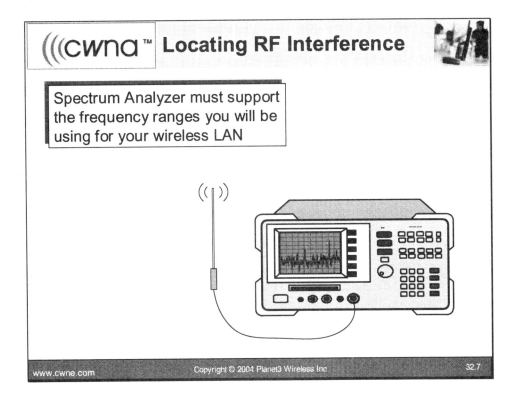

Use a spectrum analyzer (hardware or software) to locate existing sources of narrowband and spread spectrum RF, existing 802.11a/b/g installations, future installations, microwave ovens, baby monitors, spread spectrum phones, and other networks in multi-tenant buildings.

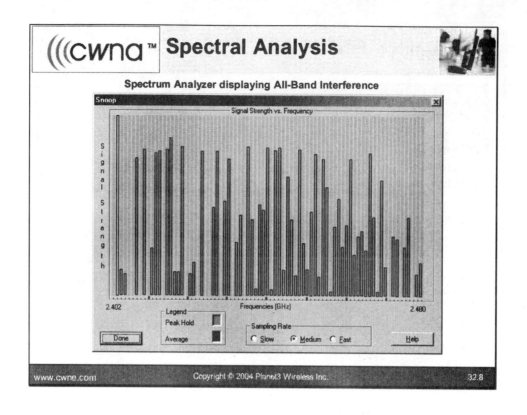
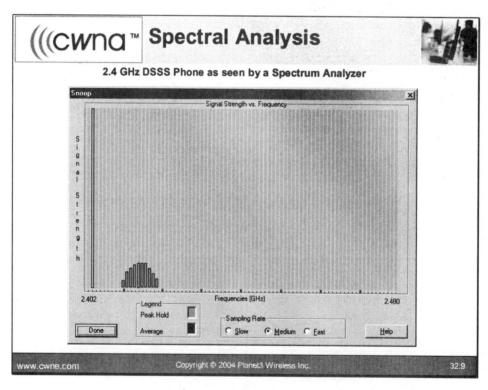

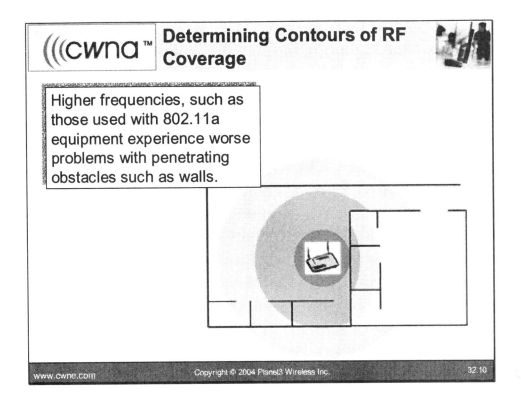

Use site surveying software to take measurements for all areas of coverage. Some wireless LAN client software includes site survey tools. Some third party companies make site survey software. Some packet analyzer software packages have site survey tools. Use the included link speed indicator to find concentric speed zones around the access point. Find & document gaps in RF coverage for particular areas. Document obstacle-induced loss (walls, windows, cubes, etc.). Test multiple antennas to find the optimum radiation pattern for a given area if needed.

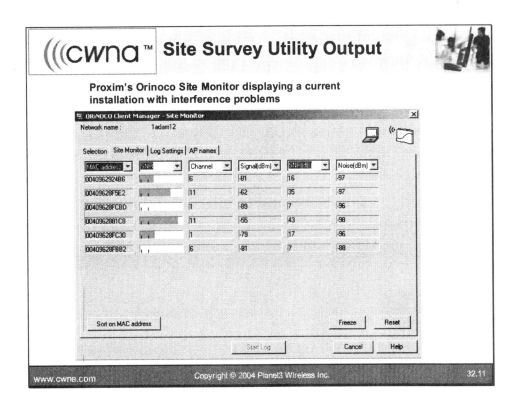

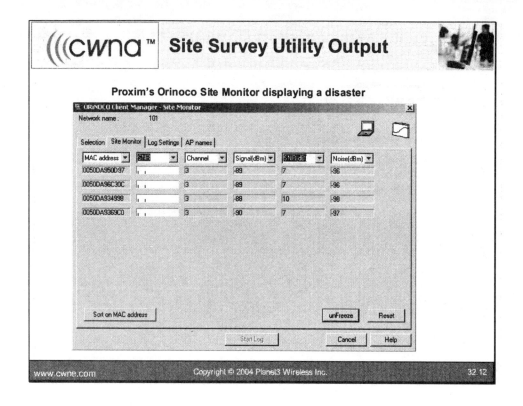

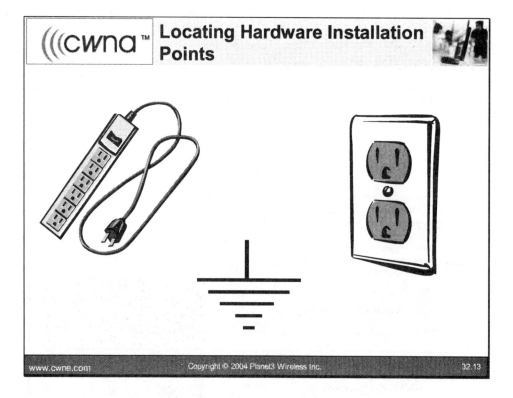

- Is AC power available near where the access point will be placed?
- Is PoE being used?
- Are wiring closets within 100 meters of any access point or bridge installation point?
- Is grounding available?

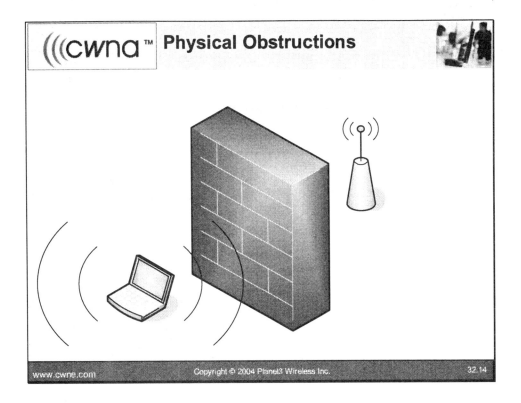

Are there physical obstructions? Firewalls or fire doors may hinder 2.4 GHz signals and stop 5 GHz signals. Pay attention to facility rules regarding fire doors, elevators, shielded rooms. Be aware of stucco walls with wire mesh and cubicles or cluttered work spaces.

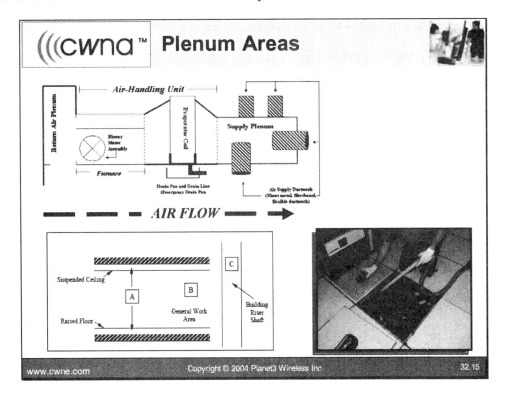

Will cabling or wireless LAN equipment be installed in the plenum or risers? Non-plenum rated cabling/equipment emits toxic fumes when burned. Suspended ceilings sometimes form part of the air distribution system and require plenum-rated cabling. Access points, antenna cabling, and data cabling should all be plenum-rated. Risers (the areas where cabling runs between floors) are surrounded by firewalls.

Copyright 2004 Planet3 Wireless, Inc.

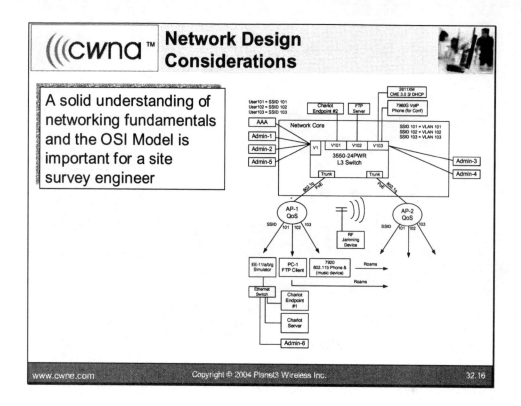

Sometimes the existing network design will dictate part or all of the wireless LAN topology. Are wired or wireless VLANs being used for segmentation? Switches were designed for stationary nodes. Mobile nodes may cross VLAN boundaries in some wireless LANs. DHCP servers are needed on each LAN segment where mobile nodes roam. Are there any routed (Layer 3) boundaries that must be crossed by mobile nodes? This may require a MobileIP solution.

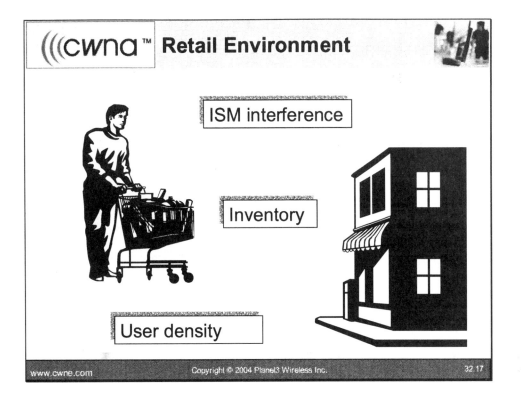

Retail has been an early adopter of wireless LAN technologies for data collection, inventory, point-of-sale, shipping/receiving. Retail may have high user density during certain periods of time like inventory. The type of traffic is usually small transactions over handheld terminals or cash registers. The store inventory causes a variable environment that must be considered.

Interference sources include:

- ISM band devices that the store sells or demos
- Baby monitors
- 2.4 GHz or 5 GHz cordless phones
- Satellite systems
- ISM band devices that the store uses in its daily business
- Cordless phones
- Neighboring stores that have ISM equipment or existing wireless LAN equipment

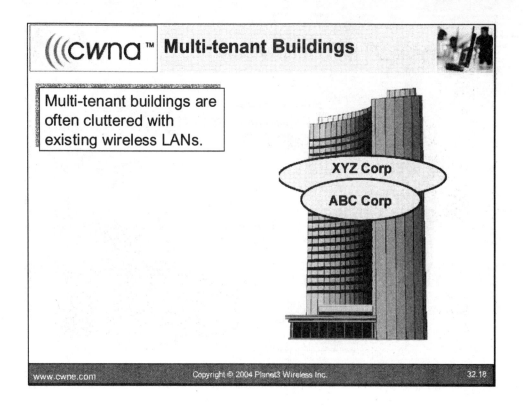

Other companies may already have wireless LANs in the building or may be planning installations. Are there any non-wireless LAN sources of 2.4 or 5 GHz interference, including elevators, cordless phones?

When surveying, try to cover multiple floors with a single access point where possible. This is possible with proper use of antennas. Multi-tenant buildings are often cluttered with existing wireless LANs. The best scenario is when building management disallows any wireless LANs except its own, which is centrally managed and provided coverage throughout.

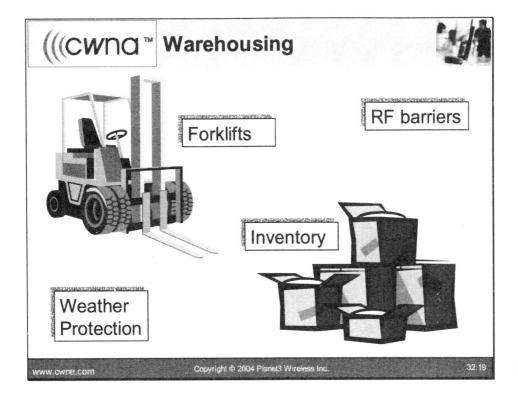

Warehouse users often prefer 2.4 GHz technologies for maximum range. Inventory types and levels present significant interference and a variable environment:

- Metal objects such as racks and devices for sale.
- Boxes of paper or paper products
- Paint and paint-related products

Consult warehouse managers and workers about inventory levels before determining RF coverage patterns. Consistently high utilization of the wireless LAN – often from many users – is common in warehousing. Mobility requirements will be much different than an office environment, requiring mobility rather than just portability. Forklifts carry computers that must maintain sessions while roaming at a fast pace. Time-sensitive applications such as Citrix may be used, requiring fast handoffs between access points.

Weather protection may be required for access points, since warehouses are not usually climate controlled. Also, warehouses are full of RF barriers:

- Chain-link fences that block off certain areas will completely block RF signals
- Shelves/Racks are often made of metal and will cause multipath and sometimes RF blockage
- Ceiling areas are typically very high in warehouses making antenna and access point mounting tricky

Different types of inventory will have different affects on RF propagation:

- Absorption
- Reflection

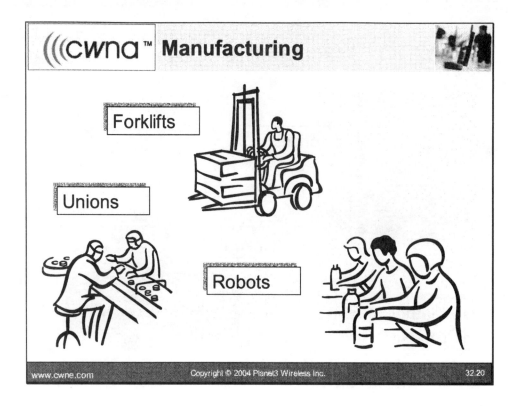

Physical damage awareness

- Do not mount access points or antennas where they are likely to be crushed by inventory or manufacturing machinery
- Be aware of robot and forklift paths

Unions

Due to large numbers of employees, there may be an employee union. Speak with the union representative about what you are allowed and not allowed to do.

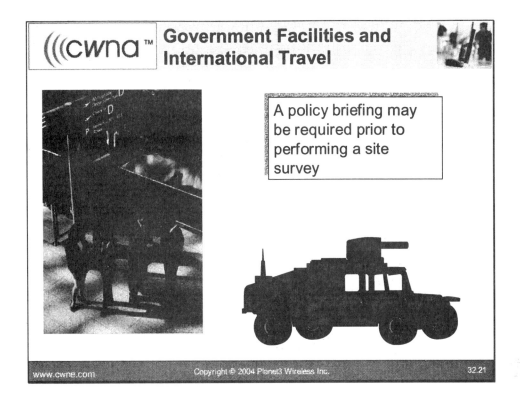

Heightened security may require that the surveyor have security clearance. Survey equipment should be cleared by security before beginning the survey. Military installations have unique requirements, so a policy briefing may be required prior to performing a site survey. A vehicle and personal search may be required prior to entering the military base. When traveling, each country's customs department will have its own rules on what is allowed inside the country. Extra paperwork and/or fees may be required both when walking the equipment in and shipping it in. A delay in customs could severely hold up the surveying project.

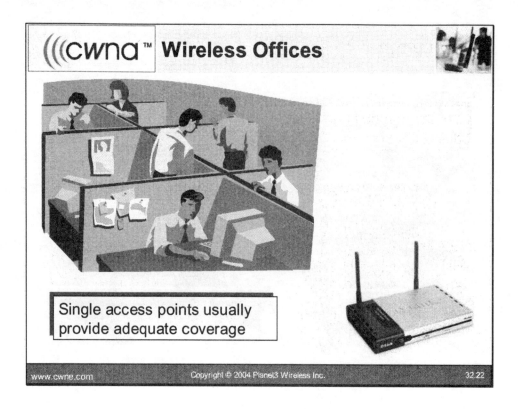

Single access points usually provide adequate coverage for small or home offices. Router/access point combination units are available from many vendors specializing in SOHO equipment:

- DLink
- Linksys
- Netgear
- Buffalo
- Microsoft
- Belkin

Some SOHO wireless routers have an integrated VPN client in order to replace three separate devices:

- Access point
- Home Internet router
- VPN Hardware Access Device

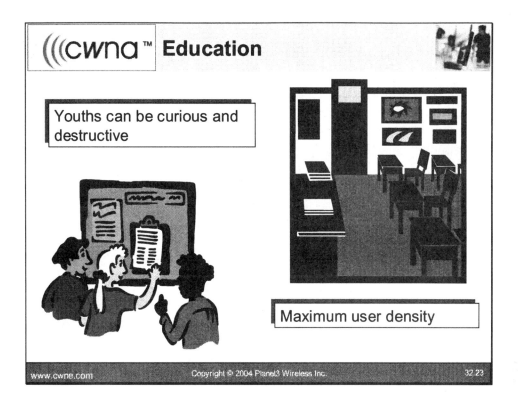

Use both 2.4 GHz and 5 GHz technologies for maximum user density in small spaces (classrooms and lecture halls). Physical security is a must at most high schools and universities, because youths can be curious and destructive. Wireless LAN equipment should be installed out of site and in lockable enclosures. Educational institutions often have Apple Macintosh computers, so a vendor must be chosen that has MAC drivers if the radio card is to be inserted into a MAC laptop. Workgroup bridges and Ethernet converters are good solutions in cases where MAC drivers are not available.

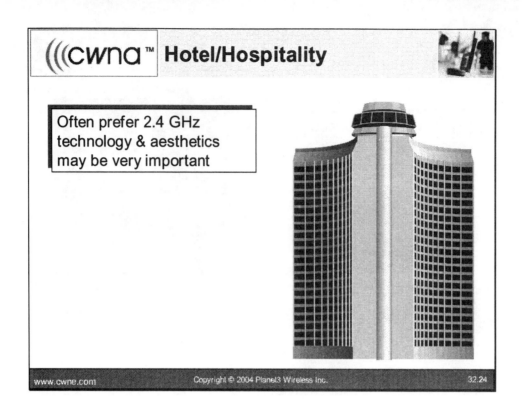

Use 2.4 GHz technologies for lowest cost while maintaining adequate speed for Internet access. When hotel rooms have wired Internet access, the lobby and mini-bars may still have wireless LAN access. Hotels having wireless LAN coverage throughout will experience some of the same issues as a multi-story office building. Physical security of wireless LAN devices is often a requirement. Aesthetics are very important to most high-end hotels. Hard cap ceilings and poured cement walls are physical obstructions to running Cat5 cable in older hotels. Most modern hotels have drop ceilings and sheet rock walls making it easier to install data cabling.

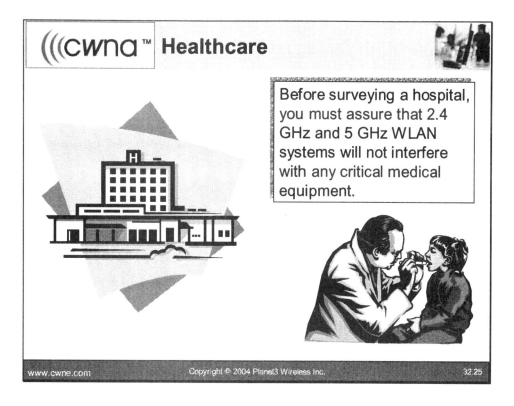

Healthcare facilities often prefer 2.4 GHz and 5 GHz technologies – each for a different use. Hospitals typically have significant 2.4 GHz ISM band gear. In those areas, 5 GHz equipment will not interfere. Healthcare facilities can be very complex to survey due to many sources of interference, multi-story buildings, long hallways, and many small rooms. Before surveying a hospital, you must assure that 2.4 GHz and 5 GHz WLAN systems will not interfere with any critical medical equipment. Most hospitals have a biomedical department that maintains all biomedical equipment. The biomedical department can certify the wireless LAN gear and site surveying tools (2-way radios) as safe to use in the hospital. Escorts are sometimes required in hospitals due to increased security. Some areas are off limits without an escort and/or an appointment, including:

- Emergency Rooms
- Operating Rooms
- Birthing Unit
- Intensive Care Unit
- Psychiatric or Criminal Wards

Appropriate customer skills and attire are often required in hospitals due to the presence of medical patients. Surveyors are often required to enter patient rooms to verify RF coverage.

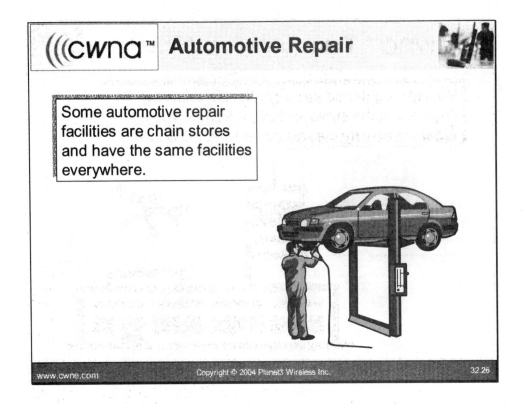

High ceilings often make custom mounting equipment necessary. If the facility is large, wireless devices that roam between access points quickly might be a requirement. Some automotive repair facilities are chain stores and have the same facilities everywhere. This makes surveying and installation quick and easy after the initial location is completed.

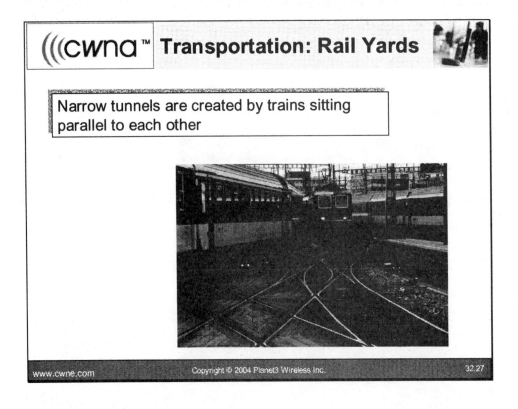

Rail cars are usually metal and transport many types of cargo that can block or reflect RF signals. Narrow tunnels or pathways are created by trains sitting parallel to each other. Yagi antennas are usually the best in these areas.

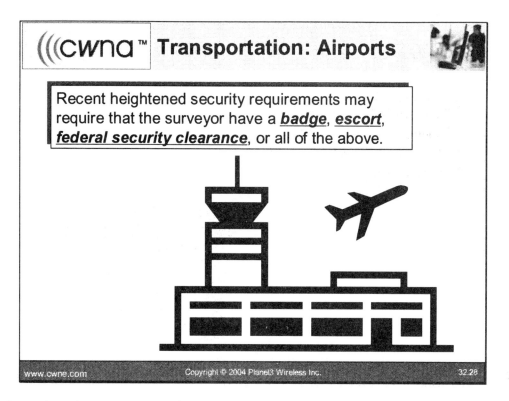

Airports always have long, open hallways with low ceilings. Baggage areas and areas around the airport facility are generally the most difficult to cover due to moving obstructions. Recent heightened security requirements may require that the surveyor have a badge, escort, federal security clearance, or all of the above. Any surveying equipment should be cleared by security before beginning the survey. Any wireless LAN installations should be within FAA and FCC regulations. While performing a survey, tools and equipment should not be left lying around in an unsecured area. Travelers could injure themselves tripping over a cable, or the equipment could be confiscated, or travelers might steal your equipment.

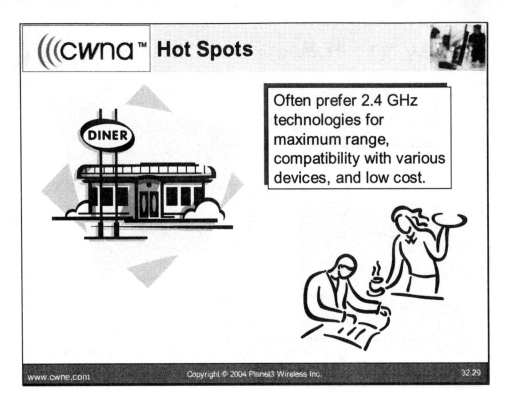

Hot Spots often prefer 2.4 GHz technologies for maximum range, compatibility with various devices, and low cost. Physical security of the wireless LAN devices is often important since users are transient. Indoor and outdoor coverage is often required. Spaces ranging from very small (coffee shop) to very large (airport) are typical.

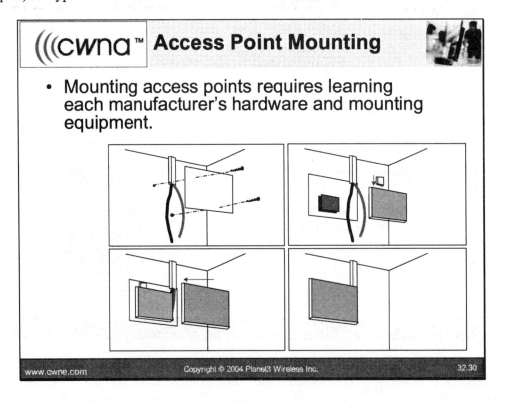

Mounting access points requires learning each manufacturer's hardware and mounting equipment. Sometimes custom mounting equipment is required.

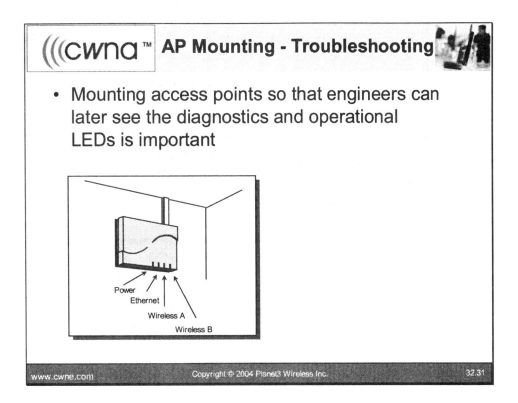

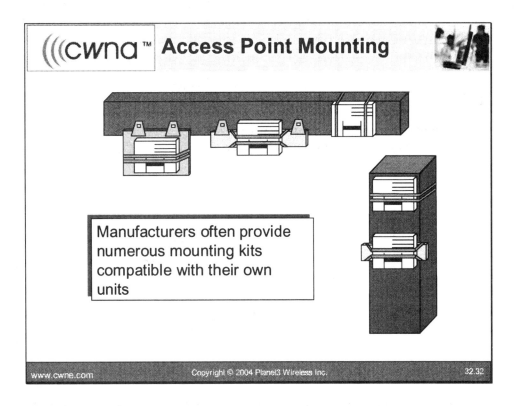

Access points can be mounted in a variety of ways. Manufacturers often provide numerous mounting kits compatible with their own units. When possible, access points should be labeled with an easily-read tag noting IP address, name, channel, and SSID.

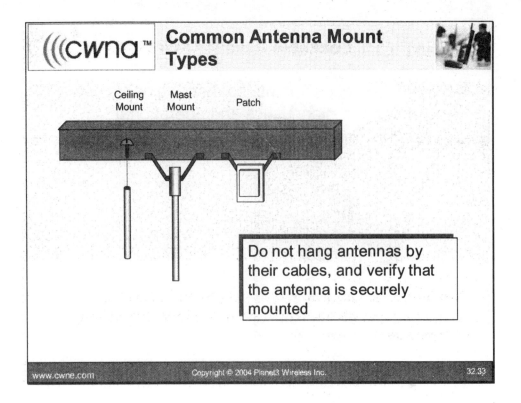

Do not hang antennas by their cables, and verify that the antenna is securely mounted. Make sure the site survey report specifies how to mount each antenna. Common antenna mount types include ceiling, wall, pillar, ground plane, mast, articulating, chimney mount, and tripod-mast.

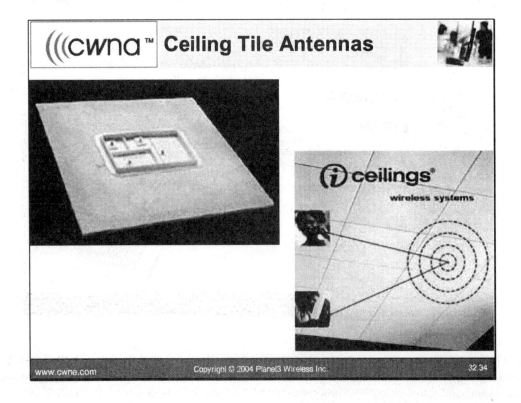

Some antennas can be hidden in plain sight. Many of these antennas support multiple wireless technologies such as 802.11, CDMA, TDMA, and GSM.

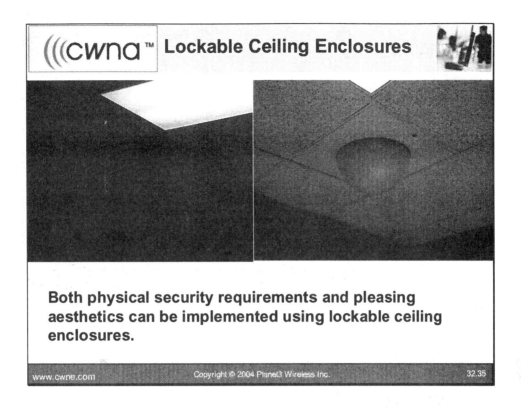

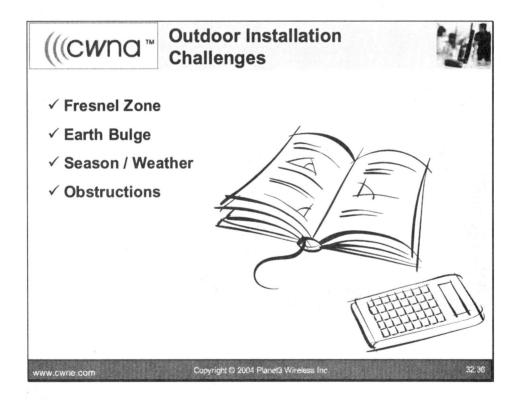

Are there trees, buildings, lakes, or other obstructions between sites? Is it winter? Trees typically allow RF to pass through in the winter. Leaves hold water when it rains, and thus, trees become a major RF obstacle. Is there RF line-of-sight between antennas? 60% clear Fresnel Zone? Is the link over 7 miles? If so, earth bulge calculation is required. Is the weather in the area volatile? Is lightning common? If so, lightning arrestors will be required. Are high winds common? If so, grid antennas may be appropriate. Is the climate rainy? If so, coax seal & NEMA enclosures will be necessary for all outdoor installations.

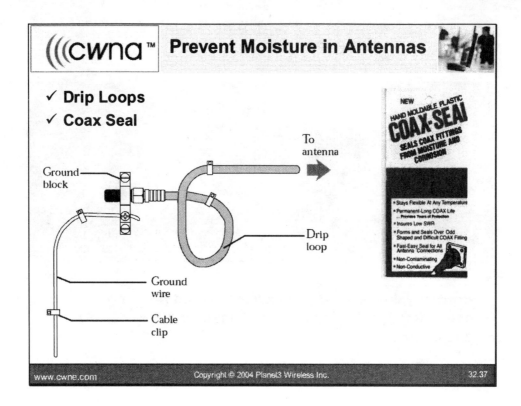

WARNING: A major problem with wireless bridge links is water in the connectors and cables causing an intermittent connection. Proper sealing is important, and Coax Seal is one product that is inexpensive and works well for sealing connectors.

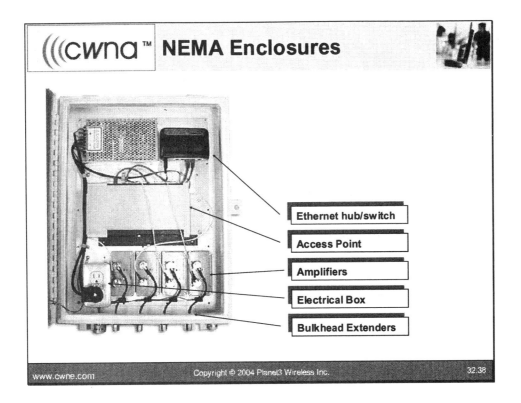

NEMA enclosures are used to house multiple wireless LAN devices outdoors in various weather conditions such as dust and moisture. NEMA enclosures are rated 1-13. NEMA Types 2, 4, & 4X typically used with wireless LAN equipment. When purchased through local electrical supply stores, enclosures are usually no more than a sealed box with no external antenna connectors, no internal mounting equipment, and no internal power supply. Bulkhead extenders will be needed to connect the access point inside to antennas mounted on the outside. Bulkhead extenders should always be on the bottom of the enclosure. Two extenders may be needed for diversity antennas.

NEMA enclosures can be purchased pre-fabricated with appropriate connectors and holes, temperature controls, and may be expensive. NEMA enclosures may require various mounting hardware. All drilled holes should be sealed with coax seal. Many WISP-oriented retailers provide enclosures as part of a package deal:

- www.ydi.com
- www.hyperlinktech.com

Towers

Is a tower required?

FCC/FAA Coordination on Antenna Towers

Antenna Tower Lighting and Marking Requirements

The FCC has been given the authority by Congress to require the painting and/or illumination of antenna towers when it determines that such towers may otherwise constitute a menace to air navigation. 47 U.S.C. § 303(q). The FCC's rules governing antenna tower lighting and painting requirements are based upon the advisory recommendations of the FAA, which are set forth in two FAA Advisory Circulars. 47 CFR §§ 17.21-17.58. Although the FAA's lighting and painting standards are advisory in nature, the FCC's rules make the standards mandatory. The standards and specifications set forth in these FAA documents are incorporated by reference into the FCC's rules, making these advisory standards mandatory for antenna towers.

The FCC always requires an FAA determination that an antenna tower will not pose an aviation hazard before it will grant permission to build that antenna tower. Information required on the FCC construction permit form advises the FCC staff as to whether such a tower location or height is involved. The FAA's determination takes into consideration the location and height of the proposed tower, and its safety lighting and marking.

Each new or altered antenna tower structure registered must conform to the FAA's painting and lighting recommendations set forth on the structure's FAA determination of "no hazard," and must be cleared with the FAA and filed with the FCC. If the FAA determines that the tower would be a physical hazard, the FCC will not approve the construction permit application. (When, however, the FAA determines that there is an aviation hazard due to possible radiofrequency interference with aviation communication signals, the FCC makes an independent analysis of who will be responsible for resolving possible conflicts, and may not automatically defer to the FAA determination as to what party should bear the cost of any needed equipment changes.)

FCC Tower Regulations: Title 47 CFR 17.7

TITLE 47--TELECOMMUNICATION

CHAPTER I--FEDERAL COMMUNICATIONS COMMISSION

PART 17--CONSTRUCTION, MARKING, AND LIGHTING OF ANTENNA STRUCTURES--Table of Contents

Subpart B--Federal Aviation Administration Notification Criteria

Sec. 17.7 Antenna structures requiring notification to the FAA.

A notification to the Federal Aviation Administration is required, except as set forth in Sec. 17.14, for any of the following construction or alteration:
 (a) Any construction or alteration of more than 60.96 meters (200 feet) in height above ground level at its site.

FCC Tower Regulations: Title 47 CFR 17.14

TITLE 47--TELECOMMUNICATION

CHAPTER I--FEDERAL COMMUNICATIONS COMMISSION

PART 17--CONSTRUCTION, MARKING, AND LIGHTING OF ANTENNA STRUCTURES--Table of Contents

Subpart B--Federal Aviation Administration Notification Criteria

Sec. 17.14 Certain antenna structures exempt from notification to the FAA.

A notification to the Federal Aviation Administration is not required for any of the following construction or alteration:
 (a) Any object that would be shielded by existing structures of a permanent and substantial character or by natural terrain or topographic features of equal or greater height, and would be located in the congested area of a city, town, or settlement where it is evident beyond all reasonable doubt that the structure so shielded will not adversely affect safety in air navigation. Applicant claiming such exemption under Sec. 17.14(a) shall submit a statement with their application to the FCC explaining basis in detail for their finding.
 (b) Any antenna structure of 6.10 meters (20 feet) or less in height except one that would increase the height of another antenna structure.
 (c) Any air navigation facility, airport visual approach or landing aid, aircraft arresting device, or meteorological device, of a type approved by the Administrator of the Federal Aviation Administration, the location and height of which is fixed by its functional purpose.

 **FCC Tower Audits**

DA 03-2411
July 23, 2003

Wireless Telecommunications Bureau Announces 60-Day Amnesty for Structures Identified in Initial Quarterly Audit of Antenna Structures

The Wireless Telecommunications Bureau (WTB) has initiated quarterly audits of antenna structures in cooperation with the National Imaging and Mapping Agency (NIMA). In the audit process NIMA adds to their obstruction database our Antenna Structure Registration (ASR) database updates and then provides back to the Commission a list of new communication tower sites from their database not found in our updates. The towers in this list may potentially require registration with the Commission. This exchange of quarterly database updates is intended to synchronize the two agencies' databases.

In the initial audit of the two agency's databases, WTB has identified 442 communications tower sites that appear to be in violation of the Commission's antenna structure registration requirements (47 CFR, Part 17). Under these Rules all antenna structures that may pose a hazard to air navigation – generally those more than 200 feet in overall height or those structures less than 200 feet but located near airports – must be studied by the Federal Aviation Administration (FAA) and registered with the Federal Communications Commission (FCC). WTB has verified that these 442 tower sites were either greater than 200 feet tall or were less than 200 feet tall, located near an airport and failed the glide slope calculation (TOWAIR). It has also been verified that these towers have not been registered with the Commission in the ASR database. A list of these tower sites is located at wireless.fcc.gov/antenna. The site will indicate by coordinates and general location the towers which may require registration.

Chapter 33
Wired Equivalent Privacy (WEP)

Objectives

Upon completion of this chapter you will be able to:

- Describe WEP operation
- Explain the intended goals of WEP
- Explain the flaws in WEP

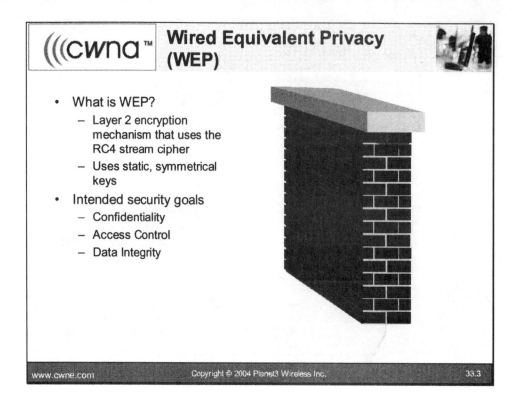

Intended goals of WEP included:

1. Confidentiality - primary goal of WEP is to prevent eavesdropping on conversations. Using WEP, data is encrypted before transmission.
2. Access Control - prevent access by unauthorized users to the network resources. Access points may be configured to reject non-encrypted or improperly encrypted frames, and stations that do not know the right WEP key won't be able to access the access point.
3. Data Integrity - prevent data from being modified, either by hackers or corruption, the checksum is computed on data before encryption.

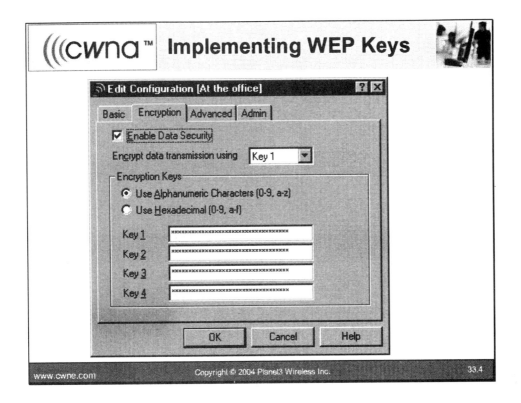

Static WEP keys can be entered in Hex or ASCII. Not all clients or access points support both Hex and ASCII. In many cases, the user can choose the default transmission key. WEP is optional.

Most clients and access points support use of up to 4 independent WEP keys. In most cases, any of the WEP keys can be chosen as the default transmission key. The keys must match exactly (in order) on each both sides of a link for encryption/decryption to work properly. A client or access point may use one key to encrypt outbound traffic and a different key to decrypt received traffic. The key used for receiving depends on which key was used by the transmitting device.

Wireless LANs were rushed to market by manufacturers, which is why WEP is not the perfect security solution. WEP satisfies the requirements of 802.11. The 802.11 standard leaves WEP implementation to vendors.

Using WEP

- Keys must match on both sides

- Protocol analyzers see "WEP Data"

- Overhead realized is vendor specific

- Switch to WPA as soon as possible

WEP secret keys must be exactly the same on each end of a connection. Secret keys only work on wireless links. When WEP is enabled, a wireless protocol analyzer will see "WEP Data" instead of the real data because WEP is a L2 protocol. The amount of throughput loss incurred due to WEP is dependent on how WEP is implemented (hardware/software).

Security Issues

Using static keys, the WEP Initialization Vector (IV) is 24-bits long and always used sequentially. In a high traffic environment, the chance is good of a WEP key being reused multiple times per day. Cracking WEP isn't feasible in most environments, but it can be done using software such as WEPcrack, or AirSnort. Cracking WEP requires large number of captured packets, long periods of time to capture enough packets, and heavy processing gear.

To avoid having your WEP cracked, use 128-bit keys when possible, and use static WEP in SOHO environments and upgrade to WPA (with TKIP) as soon as possible.

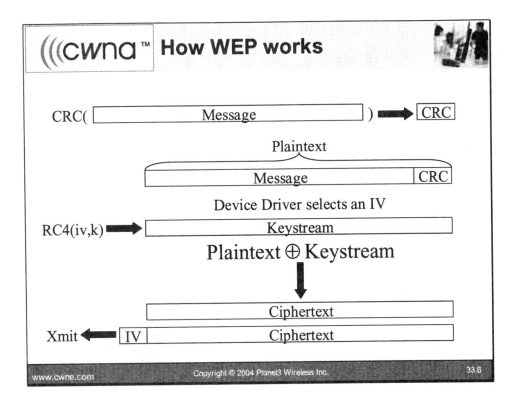

How WEP Works

A system chooses an Initialization Vector (IV), which is a 24 bit number, selected by the device drivers. Then the user or admin chooses a secret key (a.k.a. WEP Key), which is either 40-bit or 104-bit number, configured by the administrator. WEP uses these two numbers as the seeds for a random number generation algorithm. WEP generates a string of pseudo-random bytes equal in length to the plaintext to be encrypted. This is known as the keystream. The plaintext is mathematically combined with the keystream to produce the encrypted ciphertext. The plaintext is XORed with the keystream. An XOR is an extremely simple mathematical operation that can be performed quickly in hardware.

Encryption and decryption are based on the IV and the Secret key. Why does WEP have both an IV and a secret key? The secret key is hard-configured into all stations. RC4 encryption is trivial to break if you can capture multiple packets that were encrypted using the same keystream. If only the secret key were used, then all packets would be encrypted with the same keystream. The IV allows each packet to be encrypted with a different keystream. The IV is selected by the device driver, and it changes each time a packet is encrypted.

The IV is transmitted unencrypted. The IV is required to decrypt the ciphertext. The IV is selected by the device driver and is not known by the recipient. How can the recipient know the IV that was used to encrypt a packet? The IV is pre-pended – unencrypted – to the packet before transmission. Doesn't this compromise security? Since encryption is based on both the WEP key and the IV, just knowing the IV is not enough to decrypt the packet.

How the IV is selected

The 802.11 standard places no restrictions on how the IV is selected. In fact, it does not even require that the IV be changed between packets! A card could use the same IV for all packets, seriously compromising security. Typically, cards will initialize the IV to some value when they are inserted and then increment that number by one every time a packet is encrypted. Many cards initialize to zero and count up from there.

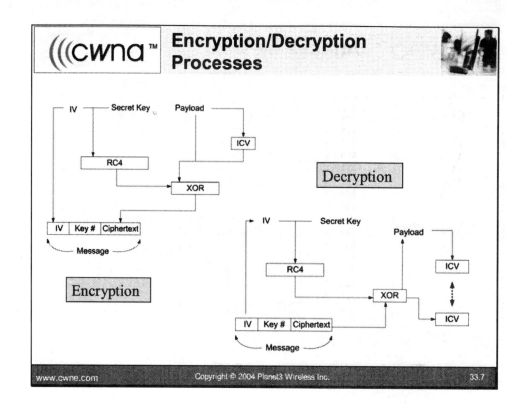
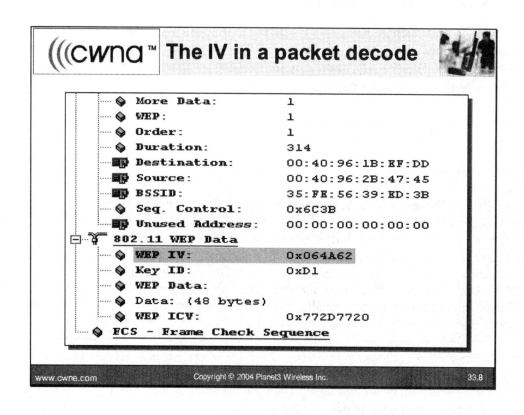

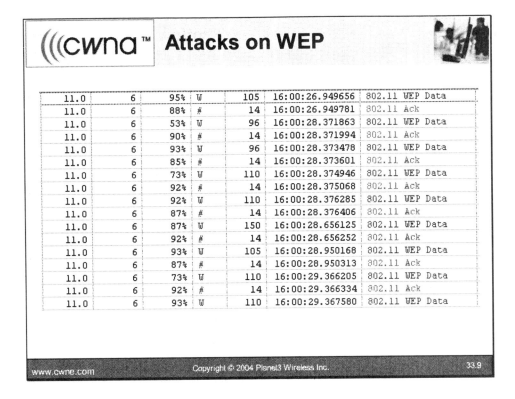

Due to flaws in the WEP algorithm, the following attacks are possible and feasible with today's technology:

- Passive attacks to decrypt traffic based on statistical analysis
- Active attacks to inject traffic from unauthorized mobile stations, based on known plaintext
- Active attacks to decrypt traffic, based on tricking the access point (into decrypting the packets for us)
- Dictionary-building attack that, after analysis of approximately a day's worth of traffic, allows real-time automated decryption of all traffic

Why WEP is weak

RC4 algorithms are only strong if the keystreams are unique. If two packets are encrypted with the same keystream, it becomes easy to break the encryption. The 802.11 standard addresses this problem with the IV. Since the IV is supposed to be different for each packet, each packet's keystream should be unique. However, IVs repeat at a rate greater than protocol designers had anticipated. When an IV is repeated (reused), it is known as an IV collision. Since the IV field is only 24 bits long, the IV must repeat at least every 2^{24} packets. This seems like a large number, but a busy access point may exhaust this address space in about five hours. Cards often initialize to the same IV when inserted. Two cards inserted at about the same time will send a multitude of packets that are encrypted with the same IV.

Sophisticated attacks do not rely on IV collisions, but rather by getting an unencrypted and encrypted copy of the same packet. Many access points send broadcast packets both encrypted and unencrypted. First, a hacker would send an email message to a user on the wireless network. By reverse-engineering that message, it is possible to decrypt other messages on the network. Analysis reveals that it is also possible to maliciously modify packets without disrupting the CRC. Hackers can change the value of arbitrary bits and fix the checksum so that it is still correct, even if they can't decrypt the packet. Combine this with the ability to decrypt the packet, and it is more dangerous, meaning that someone can change the source and destination IP address in the packets, which can change the value of commands to servers.

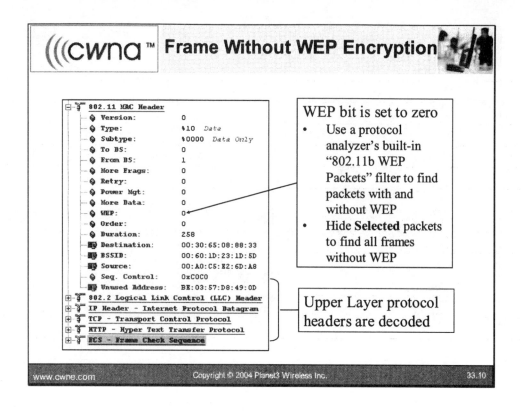

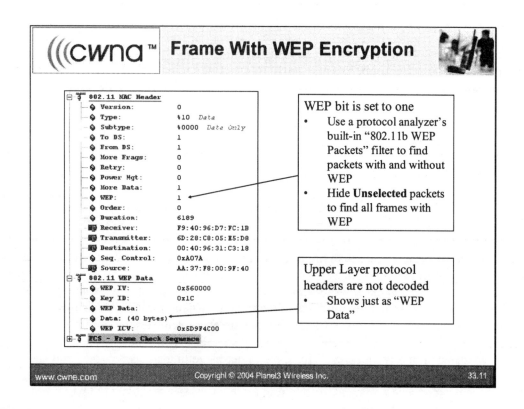

Introduction to CWNA Lab Exercises

Manufacturer's Defaults

Most of the labs in the CWNA training course require the use of one or more access points. In order to ensure successful exercises, it is necessary to reset each access point back to the manufacturer's default settings. Below are the specific instructions for the different access points utilized in these lab exercises. You may need to refer back to this page often during the course of the lab exercises.

Cisco AP-1200

If you need to start over during the initial setup process, follow these steps to reset the access point to factory default settings using the access point MODE button:

Step 1

Disconnect power (the power jack for external power or the Ethernet cable for in-line power) from the access point.

Step 2

Press and hold the MODE button while you reconnect power to the access point.

Step 3

Hold the MODE button until the Status LED turns amber (approximately 1 to 2 seconds), and release the button. All access point settings return to factory defaults.

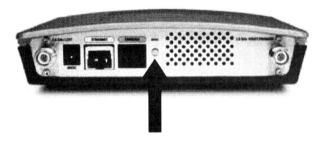

Mode Button

Proxim AP-2000

Use this procedure to reset the network configuration values, including the Access Point's IP address and subnet mask. The current AP Image is not deleted if you use this procedure. Follow this procedure if you forget the Access Point's password:

Step 1

Press and hold the RELOAD button for 10 seconds.

 Be careful to properly identify the RELOAD and RESET Buttons. You need to use a pin or the end of a paperclip to press a button. Result: The AP reboots, and the factory default network values are restored.

Step 2

If not using DHCP, use the ScanTool or CLI over a serial connection to set the IP address, subnet mask, and other IP parameters. See Command Line Interface (CLI) in the product manual for CLI information.

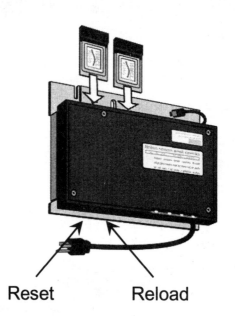

Reset   Reload

Colubris CN1250

To reset the CN1250 to its factory default settings, do the following:

Step 1

Press and hold the reset button. All the lights on the CN1250 front panel will light up.

Step 2

When the lights begin to flash (after about five seconds) immediately release the button.

Step 3

The CN1250 will restart with factory default settings. When the power light stops flashing, the CN1250 is fully operational.

Resetting the CN1250 deletes all your configuration settings, and resets the Administrator username and password to 'admin', the Wireless port IP address to 192.168.1.1, and the LAN port IP address to 192.168.4.1.

Lab 1: Infrastructure Mode Connectivity

Objectives

After completing this lab, you will be able to demonstrate how a wireless client authenticates and associates to an access point using RF technology.

Introduction

An access point maintains a table of actively connected users. This table is called the "association table" or "learning table," and contains information such as the MAC address, IP address, computer name, and status of the associated stations. The status of each client is shown in the association table (depending on the equipment vendor), and each client can be in one of the following states:

- Not authenticated and not associated
- Authenticated but not associated
- Authenticated and associated

The authentication process happens before the association process. To complete the authentication and association processes, certain settings are required at both the access point and the client stations. These settings are as follows:

- The SSID setting on each wireless client must match the SSID setting on the access point
- WEP must be configured in the same way on both ends of the connection (either off or on)
- If WEP is used, WEP keys must match exactly
- Client stations and access points must have enough power output to reach each other with their RF signal
- The access point must support and be set for the same authentication process used by the client station

802.11a Summary Information

The channel/frequency information is set on the access point only. The client will automatically set the channel/frequency to that of the access point that it authenticates to. When using 802.11a in an indoor environment, there are 8 possible channels to choose from. Not all channels are available for use in all countries.

Channel	Center Frequency (GHz)
36	5.180
40	5.200
44	5.220
48	5.240

52	5.260
56	5.280
60	5.300
64	5.320

802.11b and 802.11g Summary Information

The channel/frequency information is set on the access point only. The client will automatically set the channel/frequency to that of the access point that it authenticates to. When using 802.11b and 802.11g, there are 14 possible channels to choose from. Not all channels are available for use in all countries. Since 802.11b and 802.11g use the same channels/frequencies, care must be taken to make sure that the two networks do not use the same, or overlapping channels at the same time.

Channel	Center Frequency (GHz)	Channel	Center Frequency (GHz)
1	2.412	8	2.447
2	2.417	9	2.452
3	2.422	10	2.457
4	2.427	11	2.462
5	2.432	12	2.467
6	2.437	13	2.472
7	2.442	14	2.484

Lab 1.1: Infrastructure Mode Connectivity (802.11a)

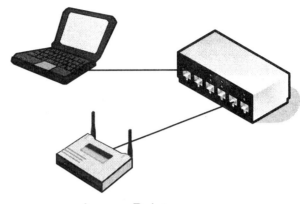

Management Station
10.0.0.2 /24

PC Card

10.0.0.100 /24
SSID=1111

Access Point
10.0.0.1 /24
Channel 36
SSID=1111

Hardware/Software Required

- 2 Laptop Computers
- 1 Access Point
- 1 wireless LAN PC Card
- 1 Ethernet switch

 Disable unused access point radios so that they will not interfere with other radios using the same frequency band in the same lab exercise. If student-owned laptops are used in the class instead of dedicated classroom-only laptops, be sure that personal firewall software is disabled. This applies to all labs in this course.

Procedure

1. Configure the network
    1.1. Connect the PC Cards to the laptop computers
    1.2. Configure hosts for a flat IP subnet (see picture for IP addresses)
2. Configure the access point
    2.1. Channel 36
    2.2. SSID=1111
    2.3. No WEP or WPA
    2.4. Open System Authentication

3. Configure wireless client
   3.1. Install and configure wireless LAN utility software
   3.2. Install and configure drivers (where required)
   3.3. SSID=1111
   3.4. No WEP or WPA
4. View the Association Table
   4.1. By viewing the access point's association (or learning) table, clients can be seen authenticating and associating to the access point. The association (or learning) table is viewed via a web browser (http) or console port in most access points.
5. Test connectivity
   5.1. Test connectivity between wireless client and the access point by pinging the access point and other wireless clients.

Troubleshooting

1. Verify WEP & SSID settings are correct on client and the access point
2. Verify drivers are properly loaded on the wireless LAN client
3. Verify wireless LAN utility software is installed properly on the client PC
4. Check the *association* (or *learning*) table in the access point to find the status of a particular wireless LAN client
5. Verify client is configured in "Infrastructure Mode"
6. Verify no MAC filters are configured on the access point
7. Verify radios are turned on in the access point and in the wireless LAN client

Lab 1.2: Infrastructure Mode Connectivity (802.11g)

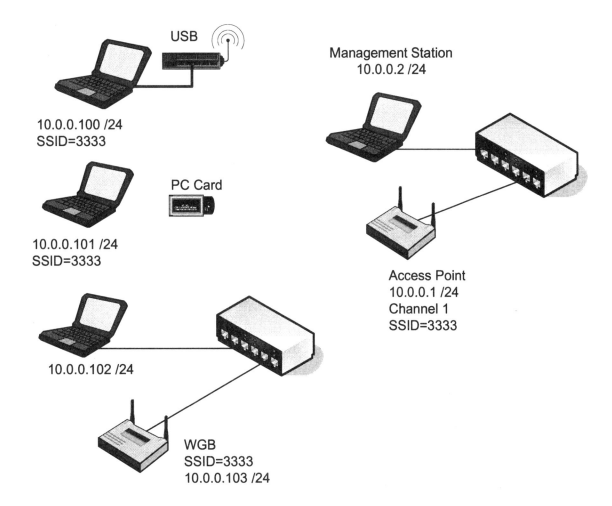

Hardware/Software Required

- 4 Laptop Computers
- 1 Access Point
- 1 wireless LAN PC Card
- 2 Ethernet switch
- 1 wireless LAN USB client
- 1 wireless LAN workgroup bridge

 Disable unused access point radios so that they will not interfere with other radios using the same frequency band in the same lab exercise.

Procedure

1. Configure the network
    1.1. Connect the PC Cards and USB Client to the laptop computers
    1.2. Configure hosts for a flat IP subnet (see picture for IP addresses)
2. Configure the access point
    - Channel 1
    - SSID=3333
    - No WEP or WPA
    - Open System Authentication
3. Configure wireless clients
    3.1. Install and configure wireless LAN utility software
    3.2. Install and configure drivers (where required)
    3.3. SSID=3333
    3.4. No WEP or WPA
4. View the Association Table
    4.1. By viewing the access point's association table, clients can be seen authenticating and associating to the access point. The association table is viewed via a web browser (http) or console port in most access points.
5. Test connectivity
    5.1. Test connectivity between wireless clients and the access point by pinging the access point and other wireless clients.

Troubleshooting

1. Verify WEP & SSID settings are correct on clients and the access point
2. Verify drivers are properly loaded on the wireless LAN clients
3. Verify wireless LAN utility software is installed properly on the client PC
4. Check the association table in the access point to find the status of a particular wireless LAN client
5. Verify clients are configured in "Infrastructure Mode"
6. Verify no MAC filters are configured on the access point
7. Verify radios are turned on in the access point and in the wireless LAN clients

Lab 2: Infrastructure Mode Throughput Analysis

Objectives

In this lab, the half-duplex nature of wireless LANs will be demonstrated. Also shown will be that frames move from wireless client to wireless client across the access point (access point relay). Use the forms provided to track and compare the throughput results.

Introduction

This lab exercise demonstrates that a wireless LAN is a half-duplex networking environment, and that frames moving from one wireless client to another wireless client must traverse the access point. These concepts are demonstrated by first moving a file using the FTP protocol between a wireless node and a wired node. During the file transfer, throughput will be monitored using the throughput-measuring software on the FTP client. The actual throughput will typically be around 50% or less of the rated bandwidth of the system in use.

When moving the same file between two wireless clients, notice that the expected throughput is further cut in half (or less). The reason for this decrease is that each frame must be sent to the access point, and then the access point must contend for network access in order to forward the frame to the destination wireless client. If the file was moving directly between wireless clients, you would expect to see approximately the same throughput as seen between wireless and wired hosts. When sending the file between the wireless host and the wired host, the frame must only traverse the wireless LAN once causing no multi-client contention for medium access.

Wireless to Wireless Copy

Client NIC Brand/Model	Server NIC Brand/Model	Access Point Brand/Model	802.11 Protocol (a/g)	Channel	Avg. Throughput

Wireless to Wired Copy

Client NIC Brand/Model	Server NIC Brand/Model	Access Point Brand/Model	802.11 Protocol (a/g)	Channel	Avg. Throughput

Lab 2.1: Infrastructure Mode Throughput Analysis (802.11a)

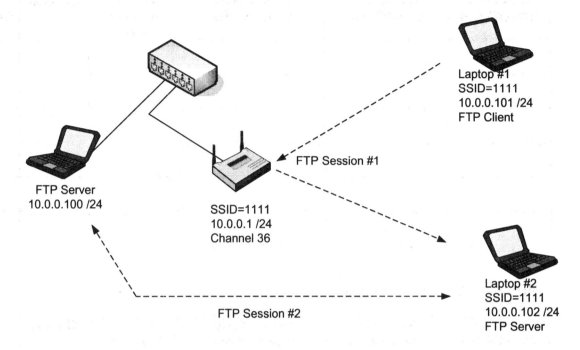

Hardware/Software Required

- 3 Laptop Computers
- 2 wireless LAN PC Cards
- 1 Access Point
- 1 Ethernet Switch
- Throughput measuring software
- FTP client software
- FTP server software

 Disable unused access point radios so that they will not interfere with other radios using the same frequency band in the same lab exercise. MakeFile.exe on the Student CD will be used to create files large enough to test throughput across high speed WLAN connections.

Procedure

1. Configure the network
    1.1. Install and configure FTP server software on the wired laptop computer
    1.2. Install and configure FTP server software on Laptop #2
    1.3. Install and configure FTP client software on Laptop #1
    1.4. Install and configure throughput measuring software on Laptop #1

1.5. Configure all hosts for a flat IP subnet (see picture for IP addresses)
2. Configure the Access Point
    2.1. Channel 36
    2.2. SSID=1111
    2.3. No WEP or WPA
    2.4. Open System Authentication
3. Configure wireless clients
    3.1. Install and configure wireless LAN utility software
    3.2. Install and configure drivers (where required)
    3.3. SSID=1111
    3.4. No WEP or WPA
4. First file transfer
    4.1. Start file transfer from Laptop #1 to the wired laptop computer
    4.2. Record average throughput on the form
5. Second file transfer
    5.1. Start file transfer from Laptop #1 to Laptop #2
    5.2. Record average throughput on the form
6. Analyze differences between file transfers
    6.1. Note the wireless-to-wireless transfer has approximately half of the throughput of the wireless-to-wired transfer

Troubleshooting

1. Verify WEP & SSID settings are correct
2. Verify drivers are properly loaded on the wireless LAN clients (where required)
3. Verify wireless LAN utility software is installed properly on the client PC
4. Check association table in access point to find status of a particular wireless LAN client
5. Verify clients are configured in "Infrastructure Mode"
6. Verify no MAC filters are configured on the access point
7. Verify radios are turned on in the access point and in the wireless LAN clients
8. Verify username and password on each FTP server are correctly matched with the FTP client
9. Verify connectivity between all devices using PING

Lab 2.2: Infrastructure Mode Throughput Analysis (802.11g)

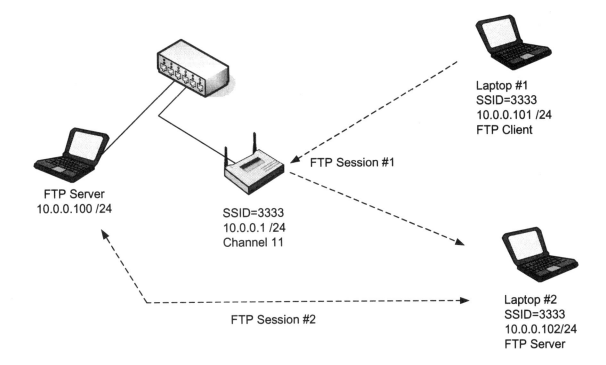

Hardware/Software Required

- 3 Laptop Computers
- 2 wireless LAN PC Cards
- 1 Access Point
- 1 Ethernet Switch
- Throughput measuring software
- FTP client software
- FTP server software

 Disable unused access point radios so that they will not interfere with other radios using the same frequency band in the same lab exercise.

Procedure

1. Configure the network
    1.1. Install and configure FTP server software on the wired laptop computer
    1.2. Install and configure FTP server software on Laptop #2
    1.3. Install and configure FTP client software on Laptop #1
    1.4. Install and configure throughput-measuring software on Laptop #1
    1.5. Configure all hosts for a flat IP subnet (see picture for IP addresses)

2. Configure the Access Point
   2.1. Channel 11
   2.2. SSID=3333
   2.3. No WEP or WPA
   2.4. Open System Authentication
3. Configure wireless clients
   3.1. Install and configure wireless LAN utility software
   3.2. Install and configure drivers (where required)
   3.3. SSID=3333
   3.4. No WEP or WPA
4. First file transfer
   4.1. Start file transfer from Laptop #1 to the wired laptop computer
   4.2. Record average throughput on the form
5. Second file transfer
   5.1. Start file transfer from Laptop #1 to Laptop #2
   5.2. Record average throughput on the form
6. Analyze differences between file transfers
   6.1. Note the wireless-to-wireless transfer has approximately half of the throughput of the wireless-to-wired transfer

Troubleshooting

1. Verify WEP & SSID settings are correct
2. Verify drivers are properly loaded on the wireless LAN clients (where required)
3. Verify wireless LAN utility software is installed properly on the client PC
4. Check association table in access point to find status of a particular wireless LAN client
5. Verify clients are configured in "Infrastructure Mode"
6. Verify no MAC filters are configured on the access point
7. Verify radios are turned on in the access point and in the wireless LAN clients
8. Verify username and password on each FTP server are correctly matched with the FTP client
9. Verify connectivity between all devices using PING

Lab 3: Ad Hoc Connectivity & Throughput Analysis

Objectives

In this lab, wireless LAN clients will connect to each other without an access point, and throughput will be calculated. This lab will be performed using the same manufacturer's hardware, and will be attempted with different manufacturer's hardware.

Introduction

An Ad Hoc network is configured using two or more wireless LAN clients and no access point. The clients must be configured on the same channel using the same SSID. The WEP, authentication, and mode settings must also match for the clients to associate (connect) with one another. The reason for defining the channel on the clients is that there are no access point broadcasting beacons, which notify clients of channel information. Some vendors offer multiple Ad Hoc modes (both standard and proprietary) and some may be incompatible with others. Some operating systems may not allow the channel to be configured in Ad Hoc mode. In this case, a different operating system that allows channel configuration in Ad Hoc mode must be used in conjunction with the one that does not.

Once both clients are fully configured with "vendor A" PC cards, check for basic connectivity with the PING utility on either side of the connection. Transfer a file using the FTP protocol between the laptop computers. During the file transfer, monitor the throughput measuring software on the FTP client. Record the throughput between wireless LAN clients. It is likely that the throughput is less than the 50% or less typically seen in Infrastructure mode using an access point.

Reconfigure one of the clients with the "vendor B" PC card and follow each test a second time. This time, it is likely that your throughput has varied enough to notice depending on the PC card manufacturers chosen.

Lab 3.1: Ad Hoc Connectivity & Throughput Analysis (802.11a)

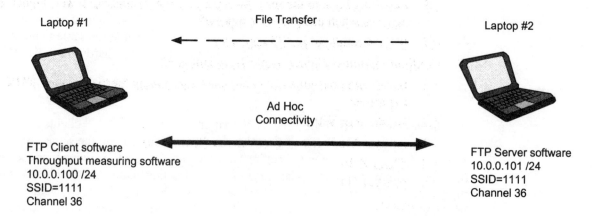

Hardware/Software Required

- 2 wireless LAN PC Cards (vendor A)
- 1 wireless LAN PC Card (vendor B)
- 2 laptop computers
- FTP Server software
- FTP Client software
- Throughput measuring software

 Disable unused access point radios so that they will not interfere with other radios using the same frequency band in the same lab exercise.

Procedure

1. Configure the network
    1.1. Install and configure FTP server software on Laptop #2
    1.2. Install and configure FTP client software on Laptop #1
    1.3. Install and configure throughput-measuring software on Laptop #1
    1.4. Configure both hosts for a flat IP subnet (see picture for IP addresses)
2. Configure wireless clients (vendor A)
    2.1. Install and configure utility software and drivers for "vendor A" PC cards in both computers
    2.2. No WEP or WPA
    2.3. Ad Hoc Mode (also called Peer-to-Peer mode)
    2.4. Channel 36
    2.5. SSID=1111
3. File transfer

- 3.1. Start file transfer from Laptop #2 to Laptop #1
- 3.2. Record average throughput
4. Analyze file transfer
    - 4.1. Referring back to the form from lab 2, is the throughput lower, higher, or the same as when using an access point?
    - 4.2. If the throughput is different, why?
5. Configure wireless client (vendor B) in laptop #2
    - 5.1. Install and configure utility software and drivers for the "vendor B" PC card in Laptop #2
    - 5.2. No WEP or WPA
    - 5.3. Ad Hoc Mode (also called Peer-to-Peer mode)
    - 5.4. Channel 36
    - 5.5. SSID=1111
6. File transfer
    - 6.1. Start file transfer from Laptop #2 to Laptop #1
    - 6.2. Record average throughput
7. Analyze File transfer
    - 7.1. Were the computers able to communicate using different vendor cards?
    - 7.2. If so, was this file transfer faster or slower than with just vendor A cards?
    - 7.3. If other 802.11a cards are available, additional testing can be performed.

Troubleshooting

1. Verify WEP & SSID settings are the same on both laptops
2. Verify drivers are properly loaded on the wireless LAN clients
3. Verify wireless LAN utility software is installed properly on the client PC
4. Verify clients are configured in "Ad Hoc Mode"
5. Verify radios are turned on in the wireless LAN clients
6. Verify username and password on the FTP server is correctly matched with the FTP client
7. Verify connectivity between devices using PING

Lab 3.2: Ad Hoc Connectivity & Throughput Analysis (802.11g)

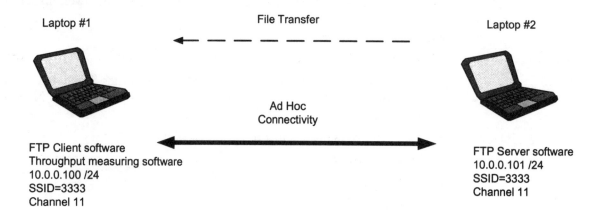

Hardware/Software Required

- 2 wireless LAN PC Cards (vendor A)
- 1 wireless LAN PC Card (vendor B)
- 2 laptop computers
- FTP Server software
- FTP Client software
- Throughput measuring software

 Disable unused access point radios so that they will not interfere with other radios using the same frequency band in the same lab exercise.

Procedure

1. Configure the network
    1.1. Install and configure FTP server software on Laptop #2
    1.2. Install and configure FTP client software on Laptop #1
    1.3. Install and configure throughput-measuring software on Laptop #1
    1.4. Configure both hosts for a flat IP subnet (see picture for IP addresses)
2. Configure wireless clients (vendor A)
    2.1. Install and configure utility software and drivers for "vendor A" PC cards in both computers
    2.2. No WEP or WPA
    2.3. Ad Hoc Mode (also called Peer-to-Peer mode)
    2.4. Channel 11
    2.5. SSID=3333
3. File transfer

3.1. Start file transfer from Laptop #1 to Laptop #2
   3.2. Record average throughput
4. Analyze file transfer
   4.1. Referring back to the form from lab 2, is the throughput lower, higher, or the same as when using an access point?
   4.2. If the throughput is different, why?
   4.3. If other 802.11g cards are available, additional testing can be performed.
5. Configure wireless clients (vendor B) in computer #2
   5.1. Install and configure utility software and drivers for the "vendor B" PC cards in Laptop #2
   5.2. No WEP or WPA
   5.3. Ad Hoc Mode (also called Peer-to-Peer mode)
   5.4. Channel 11
   5.5. SSID=3333
6. File transfer
   6.1. Start file transfer from Laptop #1 to Laptop #2
   6.2. Record average throughput
7. Analyze File transfer
   7.1. Were the computers able to communicate using different vendor cards?
   7.2. If so, was this file transfer faster or slower than with just vendor A cards?
   7.3. If other 802.11g cards are available, additional testing can be performed.

Troubleshooting

1. Verify WEP & SSID settings are the same on both laptops
2. Verify drivers are properly loaded on the wireless LAN clients
3. Verify wireless LAN utility software is installed properly on the client PC
4. Verify clients are configured in "Ad Hoc Mode"
5. Verify radios are turned on in the wireless LAN clients
6. Verify username and password on the FTP server is correctly matched with the FTP client
7. Verify connectivity between devices using PING

Lab 4: Cell-sizing and Automatic Rate Selection (ARS) in Infrastructure Mode

Objectives

In this lab exercise, cell sizing and ARS will be demonstrated. Cell sizing is important for seamless connectivity while roaming and for security purposes. ARS is the wireless LAN client's ability to speed up or slow down the wireless connection in order to maintain optimum connectivity with the access point as the client moves closer to and further from the access point. Additionally, this lab will show the effects that items in the environment can have on wireless LAN performance.

Introduction

This lab exercise will demonstrate how output power and radio sensitivity affects the size of the concentric zones of coverage within a radio cell. Each coverage zone represents slower connectivity as a client moves away from the access point. Wireless LAN client devices are designed to adapt to higher or lower rates as dictated by the received signal strength, bit error rate, and signal to noise ratio. The signal strength is based primarily on distance from the access point, but can also be affected by RF noise and interference.

Notice that while walking away from the access point, the client will automatically change (lower) connectivity speed in order to maintain the connection. This functionality is commonly known as Automatic Rate Selection or Dynamic Rate Shifting. Notice also that as output power at the access point is increased, the thickness of the zones increases, even if only slightly. At higher output power settings (30mW, 100mW, etc), the zones become even more defined. Without any significant amount of data being transferred between the wireless client and the FTP server, ARS functionality will be erratic. For example, it might cause the client to go from its highest transfer speed directly to its lowest transfer speed, or even disconnect entirely. Not all access points and clients support variable output power. Some access points support variable receiver sensitivity in order to accomplish basically the same function.

While monitoring connectivity to the access point from different distances, make a point to block the wireless LAN client and access point with items in the environment such as metal blinds, fire doors, concrete walls, etc., to see the effects on the wireless link and throughput. Start a large file transfer while the laptop is near the access point, walk away from the access point, and purposefully interfere with the connection as described above. Notice and record the effects.

Notice that while the FTP session is active, moving away from the access point will show a smoother transition between coverage zones. This is due to the sampling of more frames at the client side of the connection. More samples allow the client to make a better transition between connection speeds.

Compare the results of these labs with each other. Notice that there are differences in cell coverage between 802.11a which uses 5 GHz RF signals, and 802.11g which use 2.4 GHz RF signals. In general, the higher the frequency, the shorter the RF signal travels. Additionally, 802.11a legal output power limitations are a factor in range.

802.11a Summary Information – 802.11a equipment is required to support speeds of 6, 12, and 24 Mbps. Most 802.11a equipment supports speeds of 6, 9, 12, 18, 24, 36, 48, and 54 Mbps.

802.11g Summary Information – 802.11g equipment supports speeds of 6, 9, 12, 18, 24, 36, 48, and 54 Mbps. 802.11g equipment is also backward compatible with 802.11b, thus also supporting the speeds of 1, 2, 5.5, and 11 Mbps when in 802.11b mode.

An additional site survey lab is found in the appendix. The alternate lab uses a wireless protocol analyzer called AirMagnet to perform the site survey client functions. Both lab 4 and the additional lab can be performed provided there is adequate classroom instruction time, but the AirMagnet lab is the **PREFERRED** lab due to vendor-neutrality and more comprehensive tools than any one vendor's wireless utilities.

Lab 4.1: Cell-sizing and Automatic Rate Selection (ARS) in Infrastructure Mode (802.11a)

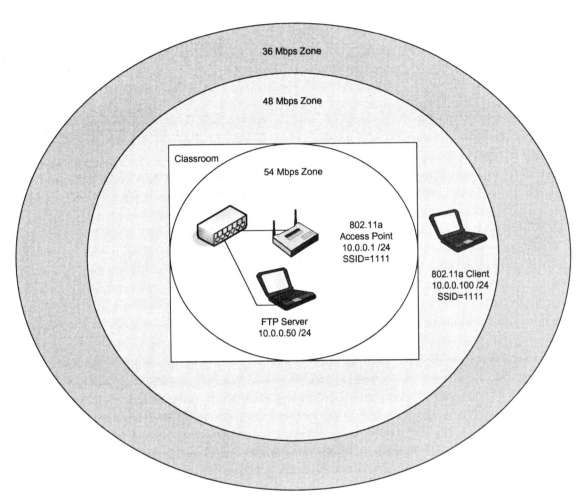

Hardware/Software required

- 1 Access Point
- 2 Laptop Computers
- 1 Ethernet Switch
- 1 wireless LAN PC Card
- FTP Server software
- FTP client software
- Throughput-measuring software
- A floor plan of the facility is helpful for determining and tracking the area of cell coverage (if available)

 Disable unused access point radios so that they will not interfere with other radios using the same frequency band in the same lab exercise.

Procedure

1. Configure the network
   1.1. Install and configure FTP server software on the wired laptop computer
   1.2. Install and configure FTP client software on the wireless laptop computer
   1.3. Install and configure throughput-measuring software on the wireless laptop computer
   1.4. Configure all hosts for a flat IP subnet (see picture for IP addresses)
2. Configure the Access Point
   - Channel 36
   - SSID=1111
   - 30 mW output power (if power level adjustment is available)
   - No WEP or WPA
   - Open System Authentication
3. Configure the wireless client
   3.1. Install and configure wireless LAN utility software
   3.2. Install and configure drivers
   3.3. 30 mW output power (if power level adjustment is available)
   3.4. SSID=1111
   3.5. No WEP or WPA
4. Testing the link
   4.1. Using the signal strength meter and other wireless LAN utilities provided by the manufacturer, determine the RF coverage area around the access point, focusing particularly on signal strength, signal-to-noise ratio, and link speed.
   4.2. Determine signal degradation caused by items in the general area around the access point such as concrete walls, metal-mesh windows, metal blinds, etc.
   4.3. If the output power of the access point and/or the client is variable, these settings can be changed to compare the differences in range between the different power settings.
5. File transfer
   5.1. Using the FTP client software, start the file transfer from the wired laptop computer to the wireless laptop computer
   5.2. Monitor throughput while moving toward and away from the access point. If a floor plan is available, mark on the floor plan where the rate shifts occur.
   5.3. Monitor throughput while putting interfering obstacles between the wireless LAN client and the access point
   5.4. Verify that the FTP session is not disrupted by the functionality of ARS

Troubleshooting

1. Verify WEP & SSID settings are the same on the laptop computers
2. Verify drivers are properly loaded on the wireless LAN client
3. Verify wireless LAN utility software is installed properly on the client PC
4. Verify the wireless LAN client is configured in "Infrastructure" mode
5. Verify the radio is turned on in the wireless LAN client
6. Verify username and password on the FTP server are correctly matched with the FTP client
7. Verify connectivity between devices using PING

3.7 mbits/sec wireless/wired

1.8 wireless/wireless

ADhoc 8.9 mbits/sec wireless/wireless

Orinoco/Netgear 8.9 mbits/sec wireless/wireless

Lab 4.2: Cell-sizing and Automatic Rate Selection (ARS) in Infrastructure Mode (802.11g)

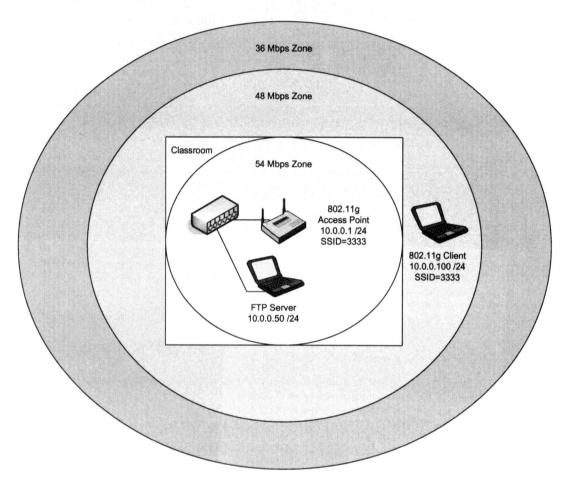

Hardware/Software required

- 1 Access Point
- 2 Laptop Computers
- 1 Ethernet Switch
- 1 wireless LAN PC Card
- FTP Server software
- FTP client software
- Throughput-measuring software
- A floor plan of the facility is helpful for determining and tracking the area of cell coverage (if available)

 Disable unused access point radios so that they will not interfere with other radios using the same frequency band in the same lab exercise.

Procedure

1. Configure the network
    1.1. Install and configure FTP server software on the wired laptop computer
    1.2. Install and configure FTP client software on the wireless laptop computer
    1.3. Install and configure throughput-measuring software on the wireless laptop computer
    1.4. Configure all hosts for a flat IP subnet (see picture for IP addresses)
2. Configure the Access Point
    2.1. Channel 11
    2.2. SSID=3333
    2.3. 30 mW output power (if power level adjustment is available)
    2.4. No WEP or WPA
    2.5. Open System Authentication
3. Configure the wireless client
    3.1. Install and configure wireless LAN utility software
    3.2. Install and configure drivers
    3.3. 30 mW output power (if power level adjustment is available)
    3.4. SSID=3333
    3.5. No WEP or WPA
4. Testing the link
    4.1. Using the signal strength meter and other wireless LAN utilities provided by the manufacturer, determine the RF coverage area around the access point, focusing particularly on signal strength, signal-to-noise ratio, and link speed
    4.2. Determine signal degradation caused by items in the general area around the access point such as concrete walls, metal-mesh windows, metal blinds, etc.
    4.3. If the output power of the access point and/or the client is variable, these settings can be changed to compare the differences in range between the different power settings.
5. File transfer
    5.1. Using the FTP client software, start the file transfer from the wired laptop computer to the wireless laptop computer
    5.2. Monitor throughput while moving toward and away from the access point. If a floor plan is available, mark on the floor plan where the rate shifts occur.
    5.3. Monitor throughput while putting interfering obstacles between the wireless LAN client and the access point
    5.4. Verify that the FTP session is not disrupted by the functionality of ARS

Troubleshooting

1. Verify WEP & SSID settings are the same on laptop and desktop computers
2. Verify drivers are properly loaded on the wireless LAN client
3. Verify wireless LAN utility software is installed properly on the client PC
4. Verify the wireless LAN client is configured in "Infrastructure" mode
5. Verify the radio is turned on in the wireless LAN client
6. Verify username and password on the FTP server are correctly matched with the FTP client
7. Verify connectivity between devices using PING

Lab 5: Co-channel & Adjacent Channel Interference

Objectives

In this lab, the effects of co-channel and adjacent channel interference on throughput, in a co-located OFDM environment will be demonstrated.

Introduction

This lab exercise shows the significant negative effects of co-channel and adjacent channel interference. The best co-location performance comes when the co-located access points have as much channel separation as possible. There are times when moving files simultaneously across the network on various channels produces catastrophic results such as practically no throughput for one of the file transfers. Another method of lessening the effects of co-channel and/or adjacent channel interference is by placing the access points further from each other or decreasing the power output of each (or both techniques simultaneously). These two techniques will significantly improve throughput by decreasing inter-access point interference.

802.11a Summary Information – Since 802.11a does not have overlapping channels, partial co-channel interference cannot be analyzed. 802.11a supports 8 non-overlapping channels for indoor use. This lab will demonstrate the effects of complete overlapping of channels and compare the effects of adjacent channel interference and separated channel interference.

802.11g Summary Information – This lab exercise shows the significant negative effects of co-channel and adjacent channel interference. This exercise will also demonstrate that the complete overlapping of channels generally produces more desirable results (better throughput) than does partial overlapping of channels. For example, the results for ch1/ch1 are better than results for ch1/ch3 (severity of degradation depends partially on environment and vendor equipment used).

Lab 5.1: Co-channel & Adjacent Channel Interference (802.11a)

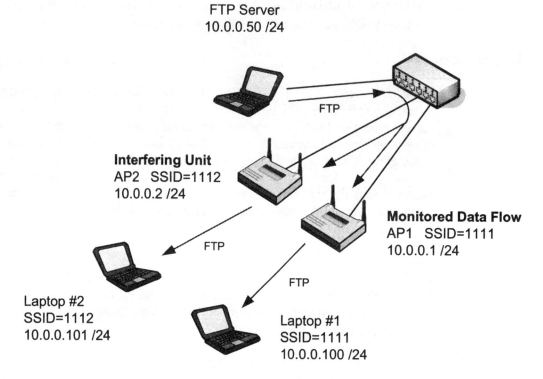

Hardware/Software Required

- 2 Access Points
- 3 Laptop computers
- 2 wireless LAN PC Cards
- 1 Ethernet Switch
- FTP Server software
- FTP client software
- Throughput-measuring software

 Disable unused access point radios so that they will not interfere with other radios using the same frequency band in the same lab exercise.

Procedure

1. Configure the network
    1.1. Install and configure FTP server software on the wired laptop computer
    1.2. Install and configure FTP client software on both wireless LAN clients
    1.3. Install and configure throughput measuring software on both wireless LAN clients
    1.4. Configure all hosts for a flat IP subnet (see picture for IP addresses)
2. Configure Access Point #1
    2.1. Channel 36
    2.2. SSID=1111
    2.3. No WEP or WPA
    2.4. Open System Authentication
3. Configure Access Point #2
    3.1. Channel 64
    3.2. SSID=1112
    3.3. No WEP or WPA
    3.4. Open System Authentication
4. Configure laptop #1
    4.1. Install and configure wireless LAN utility software
    4.2. Install and configure drivers
    4.3. SSID=1111
    4.4. No WEP or WPA
5. Configure laptop #2
    5.1. Install and configure wireless LAN utility software
    5.2. Install and configure drivers
    5.3. SSID=1112
    5.4. No WEP or WPA
6. File Transfer ch36/ch64 - non-overlapping separated channels
    6.1. Begin a file download to each of the laptops from the FTP Server simultaneously
    6.2. Monitor throughput on each laptop and record average throughput for each
7. File Transfer ch36/ch40 – non-overlapping adjacent channels
    7.1. Change Access Point #2 to channel 40
    7.2. Begin a file download to each of the laptops from the FTP Server simultaneously
    7.3. Monitor throughput on each laptop and record average throughput for each
8. File Transfer ch36/ch36 – total overlap
    8.1. Change Access Point #2 to channel 36
    8.2. Begin a file download to each of the laptops from the FTP Server simultaneously
    8.3. Monitor throughput on each laptop and record average throughput for each

Troubleshooting

1. Verify that laptop #1 is associating to Access Point #1 by checking Access Point #1's association table. If not associating, check Access Point #1 and laptop #1 settings
2. Verify that laptop #2 is associating to Access Point #2 by checking Access Point #2's association table. If not associating, check Access Point #2 and laptop #2 settings
3. Verify wireless LAN clients can ping the FTP server and have correct FTP login information
4. Verify drivers are properly loaded on the wireless LAN clients
5. Verify wireless LAN utility software is installed properly on the client PC
6. Verify wireless LAN clients are configured in "Infrastructure" mode
7. Verify radios are turned on in the wireless LAN clients and access points
8. Verify no filtering of any kind is configured on the access points
9. Verify that both access points are set to Open System Authentication with No WEP or WPA

Lab 5.2: Co-channel & Adjacent Channel Interference (802.11g)

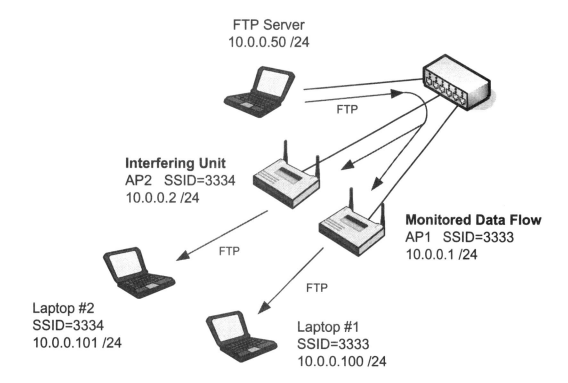

Hardware/Software Required

- 2 Access Points
- 3 Laptop computers
- 2 wireless LAN PC Cards
- 1 Ethernet Switch
- FTP Server software
- FTP client software
- Throughput-measuring software

 Disable unused access point radios so that they will not interfere with other radios using the same frequency band in the same lab exercise. This lab can also be performed using data rates up to 11 Mbps to see the effects of DSSS co-channel and adjacent channel interference.

Procedure

1. Configure the network
   1.1. Install and configure FTP server software on the wired laptop computer
   1.2. Install and configure FTP client software on both wireless LAN clients
   1.3. Install and configure throughput measuring software on both wireless LAN clients
   1.4. Configure all hosts for a flat IP subnet (see picture for IP addresses)
2. Configure Access Point #1
   2.1. Channel 1
   2.2. SSID=3333
   2.3. No WEP or WPA
   2.4. Open System Authentication
3. Configure Access Point #2
   3.1. Channel 11
   3.2. SSID=3334
   3.3. No WEP or WPA
   3.4. Open System Authentication
4. Configure laptop #1
   4.1. Install and configure wireless LAN utility software
   4.2. Install and configure drivers
   4.3. SSID=3333
   4.4. No WEP or WPA
5. Configure laptop #2
   5.1. Install and configure wireless LAN utility software
   5.2. Install and configure drivers
   5.3. SSID=3334
   5.4. No WEP or WPA
6. File Transfer ch1/ch11 - non-overlapping separated channels
   6.1. Begin a file download to each of the laptops from the FTP Server simultaneously
   6.2. Monitor throughput on each laptop and record average throughput for each
7. File Transfer ch1/ch6 – theoretically non-overlapping channels
   7.1. Change Access Point #2 to channel 6
   7.2. Begin a file download to each of the laptops from the FTP Server simultaneously
   7.3. Monitor throughput on each laptop and record average throughput for each
8. File Transfer ch1/ch3 - partially overlapping channels
   8.1. Change Access Point #2 to channel 3
   8.2. Begin a file download to each of the laptop from the FTP Server simultaneously
   8.3. Monitor throughput on each laptop and record average throughput for each

9. File Transfer ch1/ch1 – total overlap
   9.1. Change Access Point #2 to channel 1
   9.2. Begin a file transfer from each of the laptops to the FTP Server simultaneously
   9.3. Monitor throughput on each laptop and record average throughput for each

Troubleshooting

1. Verify that wireless LAN client #1 is associating to Access Point #1 by checking Access Point #1's association table. If not associating, check Access Point #1 and laptop #1 settings
2. Verify that wireless LAN client #2 is associating to Access Point #2 by checking Access Point #2's association table. If not associating, check Access Point #2 and laptop #2 settings
3. Verify wireless LAN clients can ping the FTP server and have correct FTP login information
4. Verify drivers are properly loaded on the wireless LAN clients
5. Verify wireless LAN utility software is installed properly on the client PC
6. Verify wireless LAN clients are configured in "Infrastructure" mode
7. Verify radios are turned on in the wireless LAN clients and access points
8. Verify no filtering of any kind is configured on the access points
9. Verify that both access points are set to Open System Authentication with No WEP or WPA

Lab 6: Basic Security Features

Objectives

In this lab exercise, wireless LAN client roaming between access points without losing connectivity is demonstrated. Additionally, the use of SSIDs, MAC filters, and WEP, in order to prevent unauthorized connectivity between a wireless LAN client and the access points are shown.

Introduction

This lab exercise will demonstrate roaming between access points by initially configuring both access points and the wireless LAN client with like settings. To verify roaming, installation and configuration of both access points will be done. As the wireless LAN client moves back and forth between each cell, roaming, association, and disassociation functions occur on the wireless LAN client.

Clients do not typically roam from one access point to another unless there is a need to, such as interference or loss of signal. If access points without variable output power settings are used in this lab, it may be necessary to turn off one of the access points in order to force the wireless LAN client to roam. Additionally, if this lab is performed in a relatively small space, it may be necessary to turn off one of the access points in order to force roaming regardless of output power settings on the clients and access points. A narrowband RF jamming device may be used to cause severe interference and thereby initiate roaming in the 2.4 GHz frequency space, but no cost-effective 5-6 GHz jamming devices currently exist on the open market for use with the 802.11a labs.

The next step is to block association to each of the access points using a variety of filtering methods.

- The first and most rudimentary method of blocking association is by using a different SSID, or network name. Changing the SSID of one of the access points will prevent the wireless LAN client from being able to associate to that access point.
- The second method of filtering is by enabling the MAC filter. In this lab, the MAC filter on one of the access points will be configured to demonstrate that association can be blocked using this type of filtering.
- The third method of filtering is by using WEP keys with Shared Key authentication. An access point will be configured in Shared Key authentication mode using a 64-bit (also called 40-bit) WEP key to prevent frames arriving at the access point from being forwarded.
- The fourth, and last, method of filtering is by using WPA-PSK (Wi-Fi Protected Access Pre-Shared Key). WPA-PSK is a subset of the upcoming IEEE 802.11i security standard. WPA addresses the weaknesses and vulnerabilities of WEP.

WPA is considered to be secure, however the other methods are not considered secure due to the fact that they only prevent casual eavesdropping on the wireless LAN. However, they do serve as rudimentary security measures and should be used as the most basic forms of wireless LAN security.

Lab 6.1: Basic Security Features (802.11a)

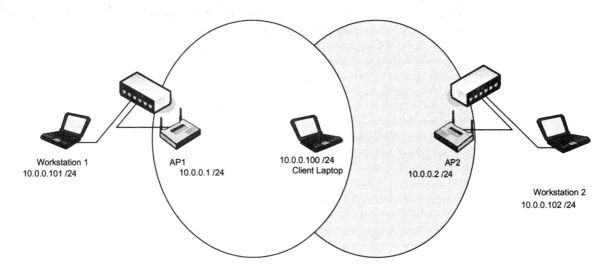

Hardware/Software required

- 2 Access Points
- 3 Laptop Computer
- 1 wireless LAN PC Card
- 2 Ethernet Switches

 Disable unused access point radios so that they will not interfere with other radios using the same frequency band in the same lab exercise.

Procedure

1. Configure the network
    1.1. Configure all devices for a flat IP subnet (see picture for IP addresses)
2. Configure the Access Points
    2.1. Channels 36 & 64
    2.2. SSID=1111
    2.3. No WEP or WPA
    2.4. Open System Authentication
3. Configure wireless LAN client
    3.1. SSID=1111
    3.2. No WEP or WPA
4. Verify seamless roaming between access points
    4.1. Verify that the wireless LAN client can roam between access points
    4.2. Verify Layer 2 connectivity by viewing each access point's association table
    4.3. Verify Layer 3 connectivity by using the PING utility
5. Change the SSID

- 5.1. Change the SSID on Access Point #1 to 9999
- 5.2. Verify that the wireless LAN client can no longer associate to Access Point #1
6. Configure a MAC filter
   - 6.1. Configure a MAC filter on Access Point #2
   - 6.2. Verify that the wireless LAN client can no longer associate to Access Point #2
7. Configure WEP with Open System or Shared Key Authentication (depending on the system in use)
   - 7.1. Change the SSID on Access Point #1 back to 1111
   - 7.2. Verify that the wireless LAN client can now associate to Access Point #1
   - 7.3. Configure Open System or Shared Key Authentication with a 40-bit WEP key of "**ABCDEFABCD**" (hex) on Access Point #1
   - 7.4. Verify that the wireless LAN client can no longer associate to Access Point #1
   - 7.5. Configure Open System or Shared Key Authentication with a 40-bit WEP key of "**ABCDEFABCD**" (hex) on the wireless LAN client
   - 7.6. Verify that the wireless LAN client can now associate to Access Point #1
8. Configure WPA with Pre-Shared Key (WPA-PSK) Authentication
   - 8.1. Disable the WEP settings on Access Point #1 and the wireless LAN client
   - 8.2. Verify that the wireless LAN client can associate to Access Point #1
   - 8.3. Configure WPA-PSK Authentication with a Pre-Shared Key of "**ABCDEFGHIJKLMNOPQRSTUVWXYZABCDEF**" (ASCII) on Access Point #1
   - 8.4. Verify that the wireless LAN client can no longer associate to Access Point #1
   - 8.5. Configure WPA-PSK Authentication with a Pre-Shared Key of "**ABCDEFGHIJKLMNOPQRSTUVWXYZABCDEF**" (ASCII) on the wireless LAN client
   - 8.6. Verify that the wireless LAN client can now associate to Access Point #1

Troubleshooting

1. Verify WEP, WPA, and SSID settings are properly set on the wireless LAN client and access points
2. Verify drivers are properly loaded on the wireless LAN client
3. Verify wireless LAN utility software is installed properly on the client PC
4. Verify wireless LAN client is configured in "Infrastructure" mode
5. Verify radio is turned on in the wireless LAN client and access points
6. Verify Layer 2 connectivity using access points' association tables
7. Verify Layer 3 connectivity using the PING utility
8. Verify proper filtering is configured on the access points
9. Verify that both access points are set to Open System Authentication with No WEP or WPA to begin this lab
10. Verify that both the access point and wireless client are set for the same authentication type (Open System, Shared Key, or WPA-PSK)

Lab 6.2: Basic Security Features (802.11g)

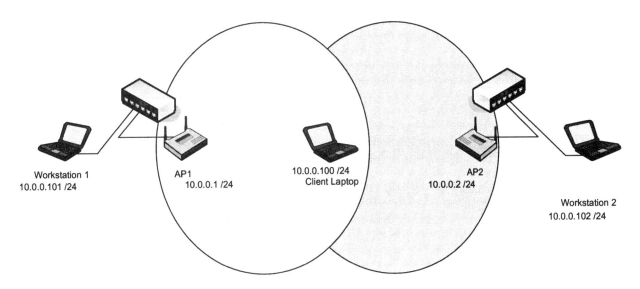

Hardware/Software required

- 2 Access Points
- 3 Laptop Computer
- 1 wireless LAN PC Card
- 2 Ethernet Switches

 If dual-radio access points are used, disable any unused radios so that they will not interfere with other radios using the same frequency band in the same lab exercise.

Procedure

1. Configure the network
    1.1. Configure all devices for a flat IP subnet (see picture for IP addresses)
2. Configure the Access Points
    2.1. Channels 1 & 11
    2.2. SSID=3333
    2.3. No WEP or WPA
    2.4. Open System Authentication
3. Configure wireless LAN client
    3.1. SSID=3333
    3.2. No WEP or WPA
4. Verify seamless roaming between access points
    4.1. Verify that the wireless LAN client can roam between access points
    4.2. Verify Layer 2 connectivity by viewing each access point's association table
    4.3. Verify Layer 3 connectivity by using the PING utility

5. Change the SSID
    5.1. Change the SSID on Access Point #1 to 9999
    5.2. Verify that the wireless LAN client can no longer associate to Access Point #1
6. Configure a MAC filter
    6.1. Configure a MAC filter on Access Point #2
    6.2. Verify that the wireless LAN client can no longer associate to Access Point #2
7. Configure WEP with Open System or Shared Key Authentication (depending on the system in use)
    7.1. Change the SSID on Access Point #1 back to 3333
    7.2. Verify that the wireless LAN client can now associate to Access Point #1
    7.3. Configure Open System or Shared Key Authentication with a 40-bit WEP key of "**ABCDEFABCD**" (hex) on Access Point #1
    7.4. Verify that the wireless LAN client can no longer associate to Access Point #1
    7.5. Configure Open System or Shared Key Authentication with a 40-bit WEP key of "**ABCDEFABCD**" (hex) on the wireless LAN client
    7.6. Verify that the wireless LAN client can now associate to Access Point #1
8. Configure WPA with Pre-Shared Key (WPA-PSK) Authentication

As of this writing, Cisco's Aironet Client Utility (ACU) does not support WPA-PSK directly. You must use Windows XP with service pack 1a and WPA patch 815485 as the WPA supplicant. In the ACU, you must choose **Select Profile → Use Another Application to Configure My Wireless Settings** in order to use the Windows XP WPA supplicant. Once WPA-PSK support is integrated into the ACU, all WPA configuration may be done from the ACU directly. The Netgear WAG511 works well as a replacement to the Cisco PC Card for this lab exercise. The WAG511 has integrated WPA-PSK support in its client utility software.

   8.1. Disable the WEP settings on Access Point #1 and the wireless LAN client
   8.2. Verify that the wireless LAN client can associate to Access Point #1
   8.3. Configure WPA-PSK Authentication with a Pre-Shared Key of "**ABCDEFGHIJKLMNOPQRSTUVWXYZABCDEF**" (ASCII) on Access Point #1
   8.4. Verify that the wireless LAN client can no longer associate to Access Point #1
   8.5. Configure WPA-PSK Authentication with a Pre-Shared Key of "**ABCDEFGHIJKLMNOPQRSTUVWXYZABCDEF**" (ASCII) on the wireless LAN client
   8.6. Verify that the wireless LAN client can now associate to Access Point #1

Troubleshooting

1. Verify WEP, WPA, and SSID settings are properly set on the wireless LAN client and access points
2. Verify drivers are properly loaded on the wireless LAN client
3. Verify wireless LAN utility software is installed properly on the client PC
4. Verify wireless LAN client is configured in "Infrastructure" mode
5. Verify radio is turned on in the wireless LAN client and access points
6. Verify Layer 2 connectivity using access points' association tables
7. Verify Layer 3 connectivity using the PING utility
8. Verify proper filtering is configured on the access points
9. Verify that both access points are set to Open System Authentication with No WEP or WPA to begin this lab
10. Verify that both the access point and wireless client are set for the same authentication type (Open System, Shared Key, or WPA-PSK)

Lab 7: Dynamic WEP keys and mutual authentication using 802.1x/EAP and RADIUS

Objectives

In this lab, a popular wireless LAN security solution is shown. This lab is intended to demonstrate the value and use of mutual authentication, dynamic WEP keys, and port-based access control (802.1x).

Introduction

This lab exercise demonstrates the use of 802.1x and the Extensible Authentication Protocol (EAP) for mutual authentication and dynamic WEP keys for a scalable wireless LAN solution. A wireless LAN card, drivers, firmware, and utility software supporting 802.1x/EAP will be installed on the client computer (laptop). The access point will support 802.1x/EAP, dynamic WEP keys, and RADIUS. The lab exercise uses both *local* RADIUS and a RADIUS server computer connected to the access point via a switch. Local RADIUS is an authentication service running inside the access point for the purpose of failover.

The RADIUS Server will be installed onto one laptop for use as the authentication server. A RADIUS server is a centralized authentication server that can generate dynamic encryption keys, verify a user's identity, and assign attributes (privileges) to a user. The access point will forward the user credentials (in this case, username and password) from the client to the RADIUS server for verification. After a successful authentication, the access point will allow the client computer onto the network. From that point forward, all frames will use WEP keys that are dynamically generated, regenerated, and distributed from the RADIUS server to the access point and client computer.

For RADIUS failover, you will have the same username and password in the Primary RADIUS server and the Secondary (local) RADIUS service. When the connection between the RADIUS server and access point is broken, clients will still be able to authenticate.

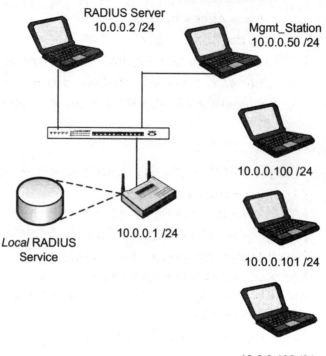

Hardware/Software Required

- 1 Cisco 1200 series Access Point
- 5 Laptop computers
- 3 Cisco wireless LAN PC Card
- 1 Ethernet Switch
- Funk Software Steel-Belted RADIUS Server software v4.5
- Cisco ACU Client software v6.2 or later

Configuration

RADIUS Server

1. Install and configure the RADIUS server as follows:
    1.1. Connect the wired laptop running RADIUS to the Ethernet switch
    1.2. Configure the IP address of the RADIUS server to 10.0.0.2 /24
    1.3. Edit the C:\RADIUS\SERVICE\EAP.INI file as follows:
        - In the section [Native-User], uncomment the first 3 lines to enable LEAP (Cisco's proprietary Lightweight EAP)
    1.4. Authentication Service Port = 1812 (by default if using Windows 2000 Pro or Windows XP Pro)
    1.5. Restart the Steel-Belted RADIUS service
    1.6. Add Access Point as RAS Client
        - Client name = AP_1

- IP Address = 10.0.0.1
- Cisco Aironet RADIUS
- Shared Secret = "key"

1.7. Create *user1* in the native RADIUS database with a password of **password**

1.8. Create *user2* in the native RADIUS database with a password of **password**

1.9. Configure the RADIUS server to check its native database first

Access Point

2. Configure LEAP on **AP** using **Mgmt_Station's** browser
    2.1. SSID = 101, Channels 1 & 36
    2.2. Primary RADIUS Server = 10.0.0.2
    2.3. Secondary (local) RADIUS Server = 10.0.0.1
    2.4. RADIUS shared secret (both servers) = *key*
    2.5. Authentication Port (both servers) = 1812
    2.6. 128-bit hex Broadcast WEP key = **0123456789ABCDEF0123456789** for Encryption key 1.
    2.7. Setup **AP** according to the following figures:

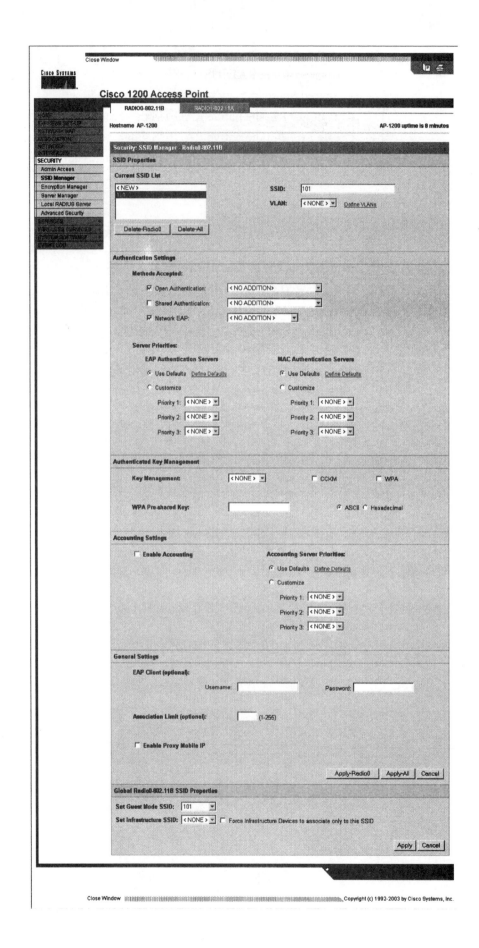

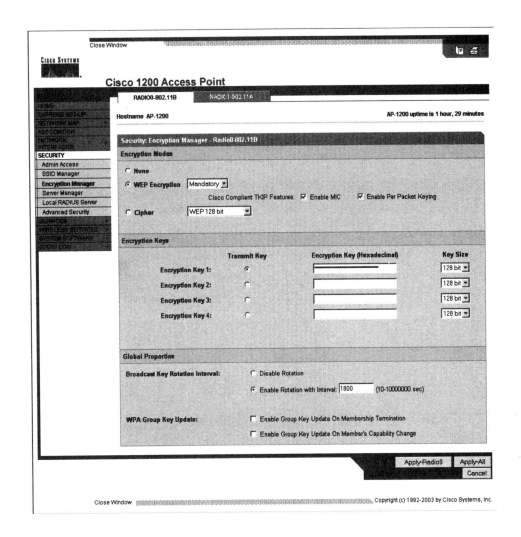

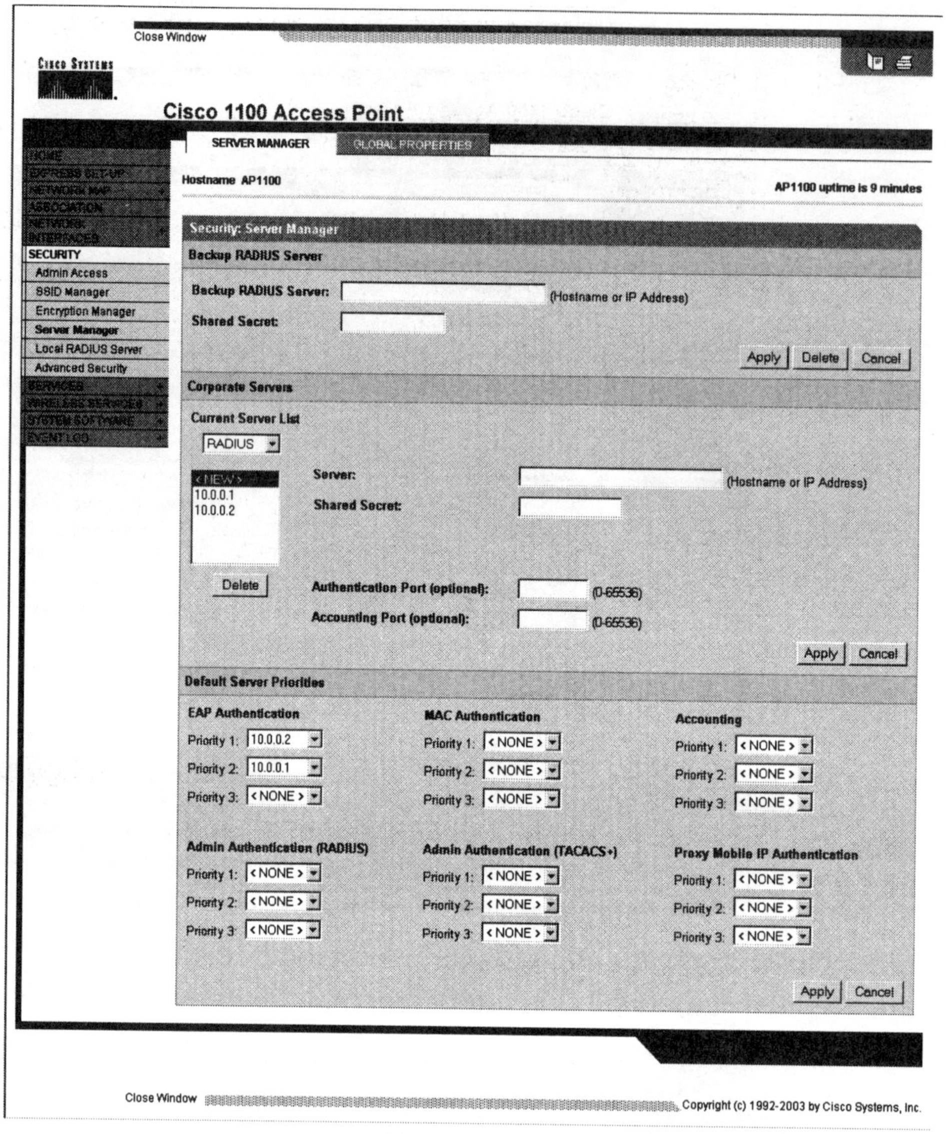
Close Window
Cisco Systems
Cisco 1100 Access Point
SERVER MANAGER
GLOBAL PROPERTIES
HOME
EXPRESS SET-UP
NETWORK MAP
ASSOCIATION
NETWORK INTERFACES
SECURITY
Admin Access
SSID Manager
Encryption Manager
Server Manager
Local RADIUS Server
Advanced Security
SERVICES
WIRELESS SERVICES
SYSTEM SOFTWARE
EVENT LOG
Hostname AP1100
AP1100 uptime is 9 minutes
Security: Server Manager
Backup RADIUS Server
Backup RADIUS Server:
(Hostname or IP Address)
Shared Secret:
Apply
Delete
Cancel
Corporate Servers
Current Server List
RADIUS
< NEW >
10.0.0.1
10.0.0.2
Server:
(Hostname or IP Address)
Shared Secret:
Delete
Authentication Port (optional):
(0-65536)
Accounting Port (optional):
(0-65536)
Apply
Cancel
Default Server Priorities
EAP Authentication
Priority 1: 10.0.0.2
Priority 2: 10.0.0.1
Priority 3: < NONE >
MAC Authentication
Priority 1: < NONE >
Priority 2: < NONE >
Priority 3: < NONE >
Accounting
Priority 1: < NONE >
Priority 2: < NONE >
Priority 3: < NONE >
Admin Authentication (RADIUS)
Priority 1: < NONE >
Priority 2: < NONE >
Priority 3: < NONE >
Admin Authentication (TACACS+)
Priority 1: < NONE >
Priority 2: < NONE >
Priority 3: < NONE >
Proxy Mobile IP Authentication
Priority 1: < NONE >
Priority 2: < NONE >
Priority 3: < NONE >
Apply
Cancel
Close Window
Copyright (c) 1992-2003 by Cisco Systems, Inc.

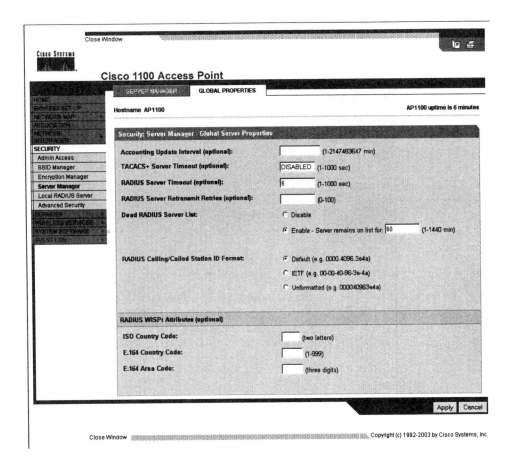

3. Enter users into *local* RADIUS database
    3.1. User name = *user1* and password = *password*
    3.2. User name = *user2* and password = *password*

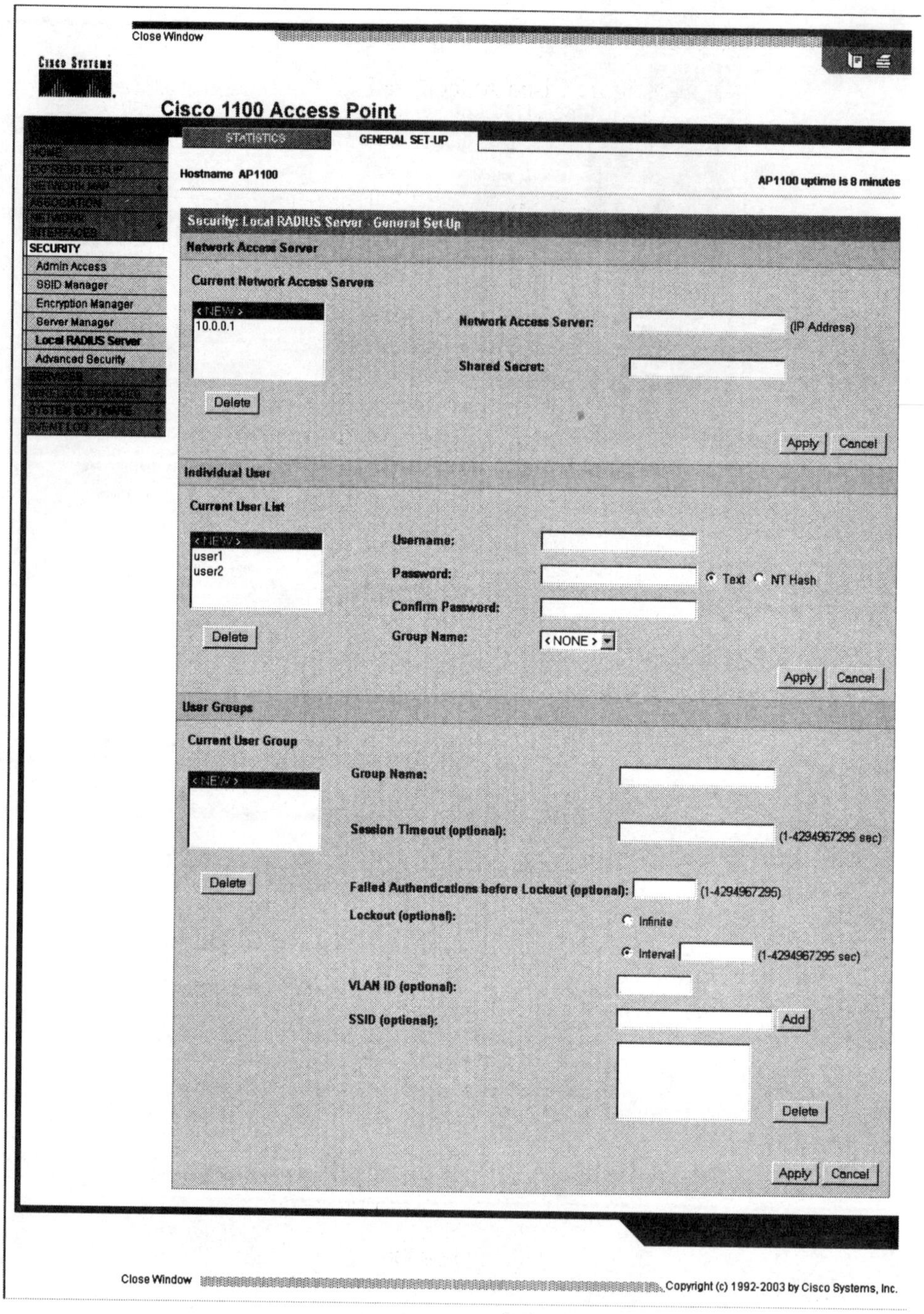
Close Window
Cisco 1100 Access Point
STATISTICS
GENERAL SET-UP
Hostname AP1100
AP1100 uptime is 8 minutes
SECURITY
Admin Access
SSID Manager
Encryption Manager
Server Manager
Local RADIUS Server
Advanced Security
Security: Local RADIUS Server - General Set-Up
Network Access Server
Current Network Access Servers
< NEW >
10.0.0.1
Network Access Server:
(IP Address)
Shared Secret:
Delete
Apply
Cancel
Individual User
Current User List
< NEW >
user1
user2
Username:
Password:
Text
NT Hash
Confirm Password:
Group Name:
< NONE >
User Groups
Current User Group
Group Name:
Session Timeout (optional):
(1-4294967295 sec)
Failed Authentications before Lockout (optional):
(1-4294967295)
Lockout (optional):
Infinite
Interval
(1-4294967295 sec)
VLAN ID (optional):
SSID (optional):
Add
Copyright (c) 1992-2003 by Cisco Systems, Inc.

Laptops

4. Configure Cisco ACU for a LEAP profile using automatic LEAP prompting for username and password. Configure an SSID of 101. Apply the LEAP profile.

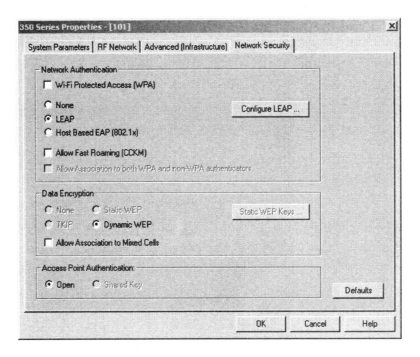

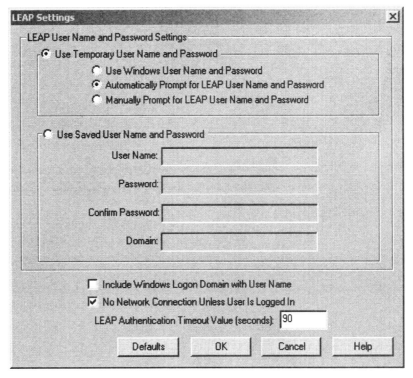

5. Verify that **Laptops** successfully associate to the **AP** using the RADIUS server for authentication.

    5.1. Prompting for authentication
    - When the client attempts authentication/association with the access point, the client software will automatically prompt the user for the username and password.

    5.2. Enter username and password for RADIUS database
    - Enter the username and password that was entered into the RADIUS server's native database. Verify that the RADIUS server has accepted the connection and verify that the client made a successful connection to the access point by pinging the RADIUS server's IP address.

6. Verify successful RADIUS server authentication by viewing the RADIUS server's *Statistics* screen, as shown below.

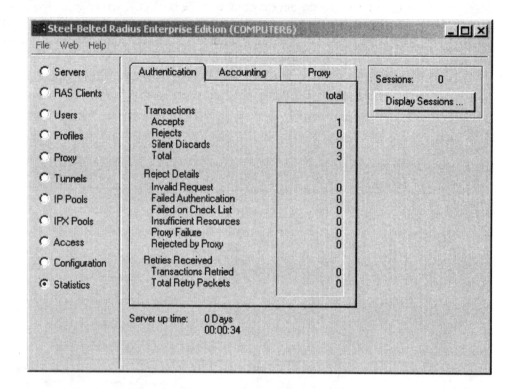

7. Disconnect the RADIUS server from the switch, and verify that laptops can still authenticate against the local RADIUS database in the AP.

Troubleshooting

1. Verify that the RADIUS server's EAP configuration is correct. This is sometimes done in a text file under the program directory. For example, you might have to verify that the appropriate LEAP section of the EAP.ini file is un-commented out. In order for this type of configuration file to be re-read by the RADIUS application, you must restart the RADIUS service on the server.

2. Verify that the RADIUS server has connected to the appropriate IP address on the server. Sometimes the server will connect to the loopback address (127.0.0.1), which will not work. If this is the case, verify your IP configuration settings on the server. This may also require that you reboot or disable/re-enable an adapter to correct the problem. When RADIUS connects to the manually configured IP address (such as 192.168.1.1), it should work fine.

3. Verify that the access point and the RADIUS server share the exact same secret key (also called a password) and IP port number (normally 1812 or 1645). The RADIUS server should have the IP address of the access point, and the access point should have the IP address of the RADIUS server.

4. Verify that you have entered the broadcast WEP key in the first slot (there are 4 slots for entering WEP keys) in the access point's WEP configuration page.

Lab 8: Wireless VPN using PPTP tunnels and RADIUS

Objectives

In this lab, a popular wireless LAN security scheme using a virtual private network (VPN) configuration is shown. The wireless LAN client will make a tunneled and encrypted connection with the VPN server (the access point in this case) and will be authenticated locally and via a RADIUS server in different steps of this procedure.

Introduction

This lab will have the client computer use PPTP VPN client software on the laptop in order to make a dial-up VPN connection with the VPN server (the access point). Current versions of Microsoft Windows support VPN services without any additional third party software. It is suggested to use this built-in VPN software, but third party software may be used provided it supports Point-to-point Tunneling Protocol (PPTP). In the first part of this lab, the VPN server will have its own native user database configured. The client computer (laptop) will dial into the IP address of the access point using an authorized username and password that has been pre-configured on the access point.

In the second part of this lab, the client will again dial into the access point, but the access point will now be configured to pass along authentication (username and password) information to a RADIUS server. The RADIUS server will verify the username and password from the client computer is valid, and notify the access point to continue building the VPN tunnel. The access point will pass an IP address to the client computer for the client end of the VPN tunnel. Once the tunnel is built, the client computer will be able to ping the RADIUS server, proving that the VPN server is now passing traffic upstream from downstream tunnels. Before the VPN tunnel is built, the client will not be able to pass traffic upstream onto the wired LAN.

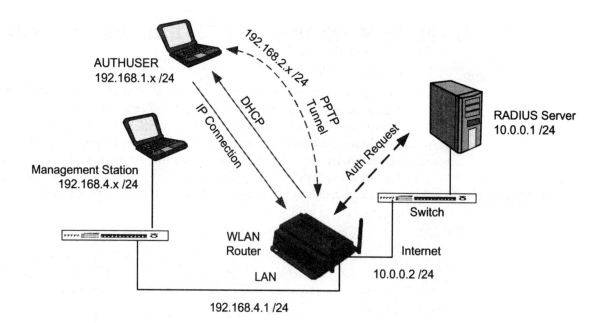

Hardware/Software Required

- 1 Colubris CN1250 WLAN Router
- 3 Laptop computers
- 1 Wireless LAN PC Card
- 2 Ethernet Switches
- Funk Software Steel-Belted RADIUS Server software v4.5

Procedure

1. Install and configure the RADIUS server
    1.1. Install and configure the RADIUS server software to take connections from the access point's IP address
    1.2. Select the appropriate RADIUS type for proper connection to the wireless router
    1.3. RADIUS key = "key"
    1.4. Connect the RADIUS server to the switch
2. Configure the Management Station
    2.1. Set the IP address to the same subnet as the access point's LAN interface
    2.2. Set the default gateway to 192.168.4.1
    2.3. Connect the management station to the switch
    2.4. Using a web browser, browse to the IP address of the WLAN Router's LAN interface
3. Configure the WLAN Router
    3.1. Reset the WLAN Router to manufacturer's defaults, then configure as follows.
    3.2. SSID = 1111
    3.3. Channel = 1

- 3.4. Security mode = PPTP
- 3.5. Authentication mode = local
- 3.6. Configure VPN tunnel IP subnet for 192.168.2.x /24
- 3.7. Configure a native user inside the WLAN Router and another (different) user in the native RADIUS database. Give each of these two users a password of "password" – a suggestion for usernames is "user1" and "user2". Configuration of users on the RADIUS server will require that the RADIUS server return two attributes for these users to the WLAN Router. These *return list attributes* will be "MS-MPPE-Recv-Key" and "MS-MPPE-Transmit-Key" and are chosen out of a list in the RADIUS server configuration.
4. Configure the client PC
    - 4.1. Configure the wireless PC Card drivers and utilities
    - 4.2. Configure the client laptop for a PPTP dial-up VPN connection to the VPN server using the WLAN Router's IP address and the username and password that were entered into the WLAN Router's local user database.
5. Use PPTP to dial the WLAN Router and verify connectivity
    - 5.1. Attempt a connection between the client PC and the WLAN Router using the username entered into the WLAN Router's local user database
    - 5.2. The client PC should establish a tunnel connection to the WLAN Router
    - 5.3. Verify proper IP address assignments by opening a console (DOS) window and typing "ipconfig". You should see a wireless adapter with your 192.168.1.x address and a VPN adapter with an IP address from the tunnel subnet (192.168.2.x)
    - 5.4. Verify proper IP connectivity by typing "ping 192.168.2.1" which should be the IP address of the WLAN Router inside the protected tunnel.
    - 5.5. Disconnect the client PC from the VPN server
6. Reconfigure Wireless Router
    - 6.1. Configure WLAN Router for "RADIUS profile 1" where the WLAN Router no longer uses its native database, but rather the external RADIUS database.
    - 6.2. Configure the WLAN Router for the external RADIUS server's IP address, port 1812, and the RADIUS key = "key"
    - 6.3. Configure the WAN interface of the WLAN Router for an IP address on the same subnet as the upstream RADIUS server.
    - 6.4. Verify that the RADIUS server can ping the WAN interface of the WLAN Router.
7. Use PPTP to dial the WLAN Router and verify connectivity
    - 7.1. Attempt a connection between the client PC and the WLAN Router using the username entered into the RADIUS server's database.
    - 7.2. The client PC should establish a tunnel connection to the WLAN Router
    - 7.3. Verify proper IP address assignments by opening a console window and typing "ipconfig". You should see a wireless adapter with your 192.168.1.x address and a VPN adapter with an IP address from the tunnel subnet (192.168.2.x)
    - 7.4. Verify proper IP connectivity by typing "ping 192.168.2.1" which should be the IP address of the WLAN Router inside the protected tunnel.
    - 7.5. Disconnect the client PC from the WLAN Router

Troubleshooting

1. Verify that the RADIUS server and WLAN Router keys match exactly
2. Verify that the RADIUS server can ping the WLAN Router
3. Verify that the WLAN Router on the client is set for PPTP
4. Verify usernames and passwords are entered into the proper places and are spelled correctly in each place
5. Verify that the WLAN Router has the latest firmware revision

Site Survey Questionnaire

Customer Name: _____

Point of Contact: _____

Address: _____

_____

Main Facility Phone Number: _____

Contact Direct Phone Number: _____

Contact Cell Phone Number: _____

Contact Email Address: _____

Contact Fax Number: _____

Interview type: ☐ In Person ☐ Phone ☐ Email

Site Survey Definition

A site survey is a task-by-task process by which the surveyor discovers and records the RF behavior throughout a facility. This information includes coverage, interference, and proper hardware placement within the facility. Site surveying involves analyzing a site from an RF perspective to determine what kind of RF coverage and hardware is required for a facility to meet the business goals of the customer.

Existing RF Network Information

Has a site survey ever been performed at this facility prior to today?

☐ YES   ☐ NO

Will any previous site surveys be made available to our staff?

☐ YES   ☐ NO

Existing Equipment at customer premises (Choose all that apply)

☐ 802.11 DSSS   ☐ 802.11B
☐ 802.11 FHSS   ☐ 802.11A
☐ 802.11 Infrared   ☐ 802.11G
☐ OpenAir FHSS   ☐ Other _____

How many existing users on the Wireless LAN? _____

How many users are there expected to be in the near future?

6 months   _____
12 months   _____
24 months   _____

Are there any peak and/or off-peak times at which certain users access the wireless LAN more than other times?

☐ YES   ☐ NO

What are the peak times?   ____ to ____ AM / PM
What are the off peak times?   ____ to ____ AM / PM

How many users does a typical access point support during peak use? _____

How many access points are currently in place? _____

What brand(s) of access points and wireless bridges are currently in place?
_____
_____
_____
_____

Are there any existing contracts for certain brands of wireless LAN hardware in place?

☐ YES   ☐ NO

Are the existing access points performing load balancing?

☐ YES   ☐ NO

What kind of Wireless LAN security solution(s) are currently in place? (Select all that apply)

☐ None	☐ Kerberos	☐ EWG
☐ WEP	☐ RADIUS	☐ EEG
☐ LEAP	☐ LDAP	☐ Firewall
☐ EAP-TLS	☐ Active Directory	☐ Router/L3 Switch
☐ EAP-TTLS	☐ NDS	☐ Other _____
☐ PEAP	☐ VLANs	
☐ VPN	☐ WVPN	

What type of wireless LAN environment is currently in place? (Choose all that apply)

☐ Campus
☐ In-building
☐ Building-to-Building
☐ MAN
☐ Other _____

Current problems with the existing wireless LAN? (Choose all that apply)

☐ Slow throughput
☐ Frequent disconnects
☐ Difficulty roaming
☐ Logon problems
☐ Other (explain thoroughly, and include documentation if available)

_____
_____
_____
_____

If the existing WLAN is having problems, has any troubleshooting of the existing problems been performed yet?

☐ YES ☐ NO

Are there any known sources of RF in or around the facility?

☐ YES ☐ NO

Source 1: _____
Source 2: _____
Source 3: _____

Are there any known RF dead zones?

☐ YES ☐ NO

If yes, attach documentation showing location(s) of dead zone(s).

Wireless LAN Administration Site Survey Questionnaire   Certified Wireless Network Professional

RF Network Design

What applications are/will be used over the wireless LAN?

☐ Voice
☐ Video
☐ Connection-oriented Data
☐ Connectionless Data
☐ Other _____

Do users roam across routed (layer 3) boundaries?

☐ YES   ☐ NO

What types of wireless client devices will be used?
☐ Laptops
☐ Desktops
☐ Handheld / PDA
☐ Scanners
☐ Thin Clients / Terminals
☐ Other _____

Are clients mounted on vehicles of any type?

☐ YES   ☐ NO

Explain _____

Is wireless printing a requirement?

☐ YES   ☐ NO

Explain _____

What network protocols and traffic types are in use over the wireless LAN?

☐ IP
☐ IPX
☐ Routing protocols
☐ Internet traffic
☐ Other _____

Are switches or hubs used for network access?

☐ Switches
☐ Hubs

Are the enough available network access ports to accommodate the access points?

☐ YES   ☐ NO

Are access points to be powered directly by AC power sources or is Power over Ethernet going to be used?

☐ AC
☐ PoE

What subnet(s) will be assigned to the wireless network?

☐ Private IP subnet
☐ Public IP subnet
☐ Subnet Information: _____

Will a naming scheme for wireless infrastructure devices be provided by the customer?

☐ YES   ☐ NO

Describe the naming conventions, and/or provide examples or naming convention documentation.
_____
_____

How will the wireless network be managed?

☐ SNMP
☐ HTTP / HTTPS
☐ Telnet
☐ Console / Serial port
☐ Custom application

Site Survey Request

What type of site survey is necessary?

☐ Indoor ☐ Outdoor

Will production application analysis required? (Without this analysis, proper application functionality cannot be guaranteed.)

☐ YES ☐ NO

Application(s): _____

If indoors, what kind of facility?

☐ Warehouse
☐ School
☐ Multi-tenant Office Building
☐ Manufacturing Plant
☐ Hospital
☐ Other _____

If outdoors, is this a point-to-point or point-to-multipoint connection? What distance(s)?

☐ Point-to-Point _____
☐ Point-to-Multipoint _____

Are there other organizations in or around your building using wireless LANs?

☐ YES ☐ NO

Need verification _____

Are Blueprints, Floor Plans, Campus Map, or other Topology Map available?

☐ YES ☐ NO

Which technology is the customer considering for the new installation? (Choose all that apply)

☐ OpenAir
☐ 802.11
☐ 802.11A
☐ 802.11B
☐ 802.11G
☐ HomeRF
☐ Other _____

Expected or Required Data rate (in Mbps)

OpenAir FHSS	☐ 0.8 ☐ 1.6
HomeRF	☐ 0.8 ☐ 1.6 ☐ 5 ☐ 10
802.11 FHSS	☐ 1 ☐ 2 ☐ Proprietary _____
802.11A	☐ 6 ☐ 9 ☐ 12 ☐ 18 ☐ 24 ☐ 36 ☐ 48 ☐ 54 ☐ Proprietary _____
802.11B	☐ 1 ☐ 2 ☐ 5.5 ☐ 11
802.11G	☐ 1 ☐ 2 ☐ 5.5 ☐ 11 ☐ 6 ☐ 9 ☐ 12 ☐ 18 ☐ 24 ☐ 36 ☐ 48 ☐ 54

☐ Other _____

Additional information :

_____
_____
_____

Special Stipulations

☐ HIPAA:
_____
_____
_____

☐ US Government
_____
_____
_____

☐ State Government:
_____
_____
_____

☐ Worker's Union
_____
_____
_____

☐ OSHA
_____
_____
_____

☐ Other:
_____
_____
_____

Site Survey – Access Point Form Certified Wireless Network Professional

Engineer Name: _____

Engineer Email: _____

Engineer Signature: _____

Customer Name: _____

Job Number: _____

AP Name / Number: _____

AP Location:

☐ Indoor ☐ Outdoor

Building / Floor: _____

Floor Plan / Map Grid Reference: _____

AP Role in Network:

☐ Root ☐ Repeater

Is more than one AP being co-located in this area for load balancing?

☐ YES ☐ NO

Wiring Closet

Name / Location _____

Data Cabling

Cable path (from AP to Wiring Closet) _____

Cable Type / Length _____

AP Type to be installed:

☐ 802.11
☐ 802.11a
☐ 802.11b
☐ 802.11g
☐ Other _____

Site Survey – Access Point Form

Certified Wireless Network Professional

Existing Network Connectivity Type:

☐ 10baseTx Hub ☐ 10/100baseTx Switch
☐ 100baseTx Hub ☐ 10/100baseFx Switch
☐ 10baseTx Switch ☐ 10/100/1000baseTx Switch
☐ 100baseTx Switch

Ethernet Switch/Hub

Mfr. / Model _____

IP Address (if managed) _____

Name / Location _____

Port Number _____

Survey Data Rate:

☐ 1 Mbps ☐ 2 Mbps ☐ 5.5 Mbps ☐ 11 Mbps
☐ 6 Mbps ☐ 9 Mbps ☐ 12 Mbps ☐ 18 Mbps
☐ 24 Mbps ☐ 36 Mbps ☐ 48 Mbps ☐ 54 Mbps
☐ Proprietary _____

Total Throughput Required from AP (if known) _____

Prevailing Traffic Types (for QoS purposes)

☐ FTP ☐ File Sharing ☐ Instant Messaging
☐ HTTP ☐ POP/SMTP ☐ Non-IP Protocols
☐ VoIP ☐ Data Backup ☐ Routing Protocols
☐ Video ☐ Database Access
☐ Telnet/SSH ☐ Warehouse Data

Types of Clients that will connect:

☐ Mobile Scanners ☐ PDA
☐ Laptop PC ☐ VoIP Phones
☐ Desktop PC ☐ Mobile Printers
☐ Other _____

Channel(s):

☐ 1 ☐ 2 ☐ 3 ☐ 4 ☐ 5 ☐ 6 ☐ 7 ☐ 8 ☐ 9 ☐ 10 ☐ 11
☐ 12 ☐ 13 ☐ 14 ☐ 36 ☐ 40 ☐ 44 ☐ 48 ☐ 52 ☐ 56 ☐ 60 ☐ 64

Site Survey – Access Point Form

Output Power:

☐ 1 mW   ☐ 2 mW   ☐ 5 mW   ☐ 10 mW   ☐ 20 mW   ☐ 32 mW
☐ 40 mW ☐ 50 mW ☐ 64 mW ☐ 100 mW ☐ 200 mW ☐ 500 mW
☐ 1 W   ☐ Other _____

AP Information

☐ AP Type/Mfr. _____

☐ Antenna Type & Gain _____

☐ Pigtail Cable _____

☐ AC Power Cabling _____

☐ Surge Protection - ☐ YES ☐ NO

☐ PoE Injector (single, multiple) _____

☐ Lightning Protection (type, ohms) _____

☐ RF Cabling & Connectors (type, length, ohms) _____

Plenum Rating Required?

☐ YES   ☐ NO

AP Housing Type

☐ NEMA Enclosure
☐ Lockable Enclosure
☐ None
☐ Other _____

AP Mounting Information

☐ Wall   ☐ Mast
☐ Ceiling   ☐ Tower
☐ Enclosure ☐ Roof
☐ Other: _____

☐ Ladder Required
☐ Lift Required
☐ Tower Climber Required

Mounting Height _____

Site Survey – Access Point Form   Certified Wireless Network Professional

Mounting Location _____

Orientation/Alignment _____

Mounting Gear Required _____

Identifying Landmarks / Items around AP _____

Notes: _____

Antenna Mounting Information

☐ Wall   ☐ Mast
☐ Ceiling   ☐ Tower
☐ Enclosure   ☐ Roof
☐ Other: _____

☐ Ladder Required
☐ Lift Required
☐ Tower Climber Required

Mounting Height _____

Mounting Location _____

Polarization: ☐ Vertical ☐ Horizontal ☐ Circular

Orientation/Alignment _____

Mounting Gear Required _____

Notes: _____

Measurement Points:

Point 1: _____
Values: Signal _____ Noise _____ SNR _____ Other _____
Interference: ☐ Narrowband ☐ WLAN ☐ ISM Equipment   ☐ Other _____
Notes: _____

Point 2: _____
Values: Signal _____ Noise _____ SNR _____ Other _____
Interference: ☐ Narrowband ☐ WLAN ☐ ISM Equipment   ☐ Other _____
Notes: _____

Site Survey – Access Point Form   Certified Wireless Network Professional

Point 3: _____
Values: Signal _____ Noise _____ SNR _____ Other _____
Interference: ☐ Narrowband ☐ WLAN ☐ ISM Equipment   ☐ Other _____
Notes: _____

Point 4: _____
Values: Signal _____ Noise _____ SNR _____ Other _____
Interference: ☐ Narrowband ☐ WLAN ☐ ISM Equipment   ☐ Other _____
Notes: _____

Point 5: _____
Values: Signal _____ Noise _____ SNR _____ Other _____
Interference: ☐ Narrowband ☐ WLAN ☐ ISM Equipment   ☐ Other _____
Notes: _____

Point 6: _____
Values: Signal _____ Noise _____ SNR _____ Other _____
Interference: ☐ Narrowband ☐ WLAN ☐ ISM Equipment   ☐ Other _____
Notes: _____

Point 7: _____
Values: Signal _____ Noise _____ SNR _____ Other _____
Interference: ☐ Narrowband ☐ WLAN ☐ ISM Equipment   ☐ Other _____
Notes: _____

Point 8: _____
Values: Signal _____ Noise _____ SNR _____ Other _____
Interference: ☐ Narrowband ☐ WLAN ☐ ISM Equipment   ☐ Other _____
Notes: _____

Point 9: _____
Values: Signal _____ Noise _____ SNR _____ Other _____
Interference: ☐ Narrowband ☐ WLAN ☐ ISM Equipment   ☐ Other _____
Notes: _____

Point 10: _____
Values: Signal _____ Noise _____ SNR _____ Other _____
Interference: ☐ Narrowband ☐ WLAN ☐ ISM Equipment   ☐ Other _____
Notes: _____

Known Dead Spots:

Point 1: _____
Caused by: _____
Suggestions: _____

Point 2: _____
Caused by: _____
Suggestions: _____

Site Survey – Access Point Form   Certified Wireless Network Professional

Point 3: _____
Caused by: _____
Suggestions: _____

Obstacles in the immediate environment:

☐ Metal blinds   ☐ Fire doors   ☐ Metal mesh windows
☐ HVAC   ☐ Duct Work   ☐ Firewall
☐ Elevator   ☐ Machinery   ☐ Warehouse shelves / goods
☐ Pipes
☐ Other _____

Environment where AP will be placed

☐ Open office space   ☐ Office with cubicles ☐ Warehouse / Distribution
☐ Retail Sales   ☐ Freezer/Cold Storage ☐ Hallway / Corridor
☐ Manufacturing   ☐ Outdoors
☐ Other _____

Elements to which the AP will be exposed

☐ Heat   ☐ Cold   ☐ Fluctuating Temperature
☐ Rain / Snow   ☐ Dirt / Dust   ☐ Grease
☐ Chemicals   ☐ Sunlight   ☐ Vibration
☐ Wind

Notes:
_____
_____
_____

Site Survey – Access Point Form   Certified Wireless Network Professional

Configuration Information

AP Management Information:

☐ HTTP _____

☐ Telnet _____

☐ SNMP _____

☐ Console/Serial Port _____

☐ Custom Application _____

Wireless VLANs:

ESSID _____ → VLAN _____   ESSID _____ → VLAN _____

ESSID _____ → VLAN _____   ESSID _____ → VLAN _____

ESSID _____ → VLAN _____   ESSID _____ → VLAN _____

ESSID _____ → VLAN _____   ESSID _____ → VLAN _____

ESSID _____ → VLAN _____   ESSID _____ → VLAN _____

ESSID _____ → VLAN _____   ESSID _____ → VLAN _____

ESSID _____ → VLAN _____   ESSID _____ → VLAN _____

ESSID _____ → VLAN _____   ESSID _____ → VLAN _____

IP Address: _____
MAC Address: _____
ESSID: _____

Authentication/Encryption:

☐ 802.1x/LEAP   ☐ 802.1x/EAP-TLS   ☐ 802.1x/PEAP   ☐ 802.1x/EAP-TTLS
☐ 802.1x/EAP-MD5   ☐ WEP Plus   ☐ PPTP VPN   ☐ IPSec VPN
☐ L2TP VPN   ☐ KeyGuard   ☐ TKIP   ☐ MIC
☐ Other _____

Authentication Server Type:

☐ RADIUS   ☐ LDAP   ☐ Active Directory/Kerberos   ☐ eDirectory/NDS
☐ Other _____

Pictures of Access Point Mounting

In the following fields, record the name and description of each digital photograph taken.

Name: _____

Description: _____

Name: _____

Description: _____

Name: _____

Description: _____

Pictures of Antenna Mounting

In the following fields, record the name and description of each digital photograph taken.

Name: _____

Description: _____

Name: _____

Description: _____

Name: _____

Description: _____

Pictures of Custom Mounting Equipment (if required)

In the following fields, record the name and description of each digital photograph taken.

Name: _____

Description: _____

Name: _____

Description: _____

Name: _____

Description: _____

Site Survey – Bridge Form　　　　　　　　　　Certified Wireless Network Professional

Engineer Name: _____

Engineer Email: _____

Engineer Signature: _____

Customer Name: _____

Job Number: _____

Bridge Name / Number: _____

Bridge Location:

☐ Indoor　　　☐ Outdoor

Building / Floor: _____

Site/Tower: _____

Map Grid Reference: _____

Bridge Role in Network:

☐ Root　　　☐ Non-Root　　　☐ Repeater

Network Architecture?

☐ Point-to-Point　　　☐ Point-to-Multipoint

Wiring Closet

Name / Location _____

Data Cabling

Cable path (from Bridge to Wiring Closet) _____

Cable Type / Length _____

Bridge Type to be installed:

☐ 802.11a
☐ 802.11b
☐ 802.11g
☐ FSO
☐ Other _____

Site Survey – Bridge Form   Certified Wireless Network Professional

UNII Bands

☐ UNII-1 (5.15-5.25 GHz)
☐ UNII-2 (5.25-5.35 GHz)
☐ UNII-3 (5.725-5.825 GHz)
☐ Licensed Frequencies _____

Existing Network Connectivity Type:

☐ 10baseTx Hub   ☐ 10/100baseTx Switch
☐ 100baseTx Hub   ☐ 10/100baseFx Switch
☐ 10baseTx Switch   ☐ 10/100/1000baseTx Switch
☐ 100baseTx Switch

Ethernet Switch/Hub

Mfr. / Model _____

IP Address (if managed) _____

Name / Location _____

Port Number _____

Survey Data Rate:

☐ 1 Mbps   ☐ 2 Mbps   ☐ 5.5 Mbps   ☐ 11 Mbps
☐ 6 Mbps   ☐ 9 Mbps   ☐ 12 Mbps   ☐ 18 Mbps
☐ 24 Mbps   ☐ 36 Mbps   ☐ 48 Mbps   ☐ 54 Mbps

☐ Proprietary _____

Total Throughput Required from Bridge (if known) _____

Will client devices connect wirelessly to the bridge?

☐ YES   ☐ NO

Prevailing Traffic Types (for QoS purposes)

☐ FTP   ☐ File Sharing   ☐ Instant Messaging
☐ HTTP   ☐ POP/SMTP   ☐ Non-IP Protocols
☐ VoIP   ☐ Data Backup   ☐ Routing Protocols
☐ Video   ☐ Database Access
☐ Telnet/SSH   ☐ Warehouse Data

Channel(s):

☐ 1   ☐ 2   ☐ 3   ☐ 4   ☐ 5   ☐ 6   ☐ 7   ☐ 8   ☐ 9   ☐ 10   ☐ 11
☐ 12   ☐ 13   ☐ 14   ☐ 36   ☐ 40   ☐ 44   ☐ 48   ☐ 52   ☐ 56   ☐ 60   ☐ 64

Site Survey – Bridge Form

Certified Wireless Network Professional

☐ 149   ☐ 153   ☐ 157   ☐ 161

Output Power:

☐ 1 mW   ☐ 2 mW   ☐ 5 mW   ☐ 10 mW   ☐ 20 mW   ☐ 32 mW
☐ 40 mW   ☐ 50 mW   ☐ 64 mW   ☐ 100 mW   ☐ 200 mW   ☐ 500 mW
☐ 1 W   ☐ Other _____

Bridge Information

☐ Bridge Type/Mfr. _____

☐ Antenna Type & Gain _____

☐ Pigtail Cable _____

☐ AC Power Cabling _____

☐ Surge Protection - ☐ YES ☐ NO

☐ PoE Injector (single, multiple) _____

☐ Lightning Protection (type, ohms) _____

☐ Grounding Information _____

☐ RF Cabling & Connectors (type, length, ohms) _____

Plenum Rating Required?

☐ YES   ☐ NO

Bridge Housing Type

☐ NEMA Enclosure
☐ Lockable Enclosure
☐ None
☐ Other _____

Bridge Mounting Information

☐ Wall   ☐ Mast
☐ Ceiling   ☐ Tower
☐ Enclosure ☐ Roof
☐ Other: _____

☐ Ladder Required
☐ Lift Required

Copyright 2004 Planet3 Wireless, Inc.

Site Survey – Bridge Form

Certified Wireless Network Professional

☐ Tower Climber Required

Mounting Height _____

Mounting Location _____

Orientation/Alignment _____

Mounting Gear Required _____

Notes: _____

Antenna Mounting Information

☐ Wall   ☐ Mast
☐ Ceiling   ☐ Tower
☐ Enclosure   ☐ Roof
☐ Other: _____

☐ Ladder Required
☐ Lift Required
☐ Tower Climber Required

Mounting Height _____

Mounting Location _____

Polarization: ☐ Vertical ☐ Horizontal ☐ Circular

Orientation/Alignment _____

Mounting Gear Required _____

Notes: _____

Measurement Points:

Point 1: _____
Values: Signal _____ Noise _____ SNR _____ Other _____
Interference: ☐ Narrowband ☐ WLAN ☐ ISM Equipment ☐ Other _____
Notes: _____

Point 2: _____
Values: Signal _____ Noise _____ SNR _____ Other _____
Interference: ☐ Narrowband ☐ WLAN ☐ ISM Equipment ☐ Other _____
Notes: _____

Point 3: _____

Site Survey – Bridge Form   Certified Wireless Network Professional

Values: Signal _____ Noise _____ SNR _____ Other _____
Interference: ☐ Narrowband ☐ WLAN ☐ ISM Equipment ☐ Other _____
Notes: _____

Point 4: _____
Values: Signal _____ Noise _____ SNR _____ Other _____
Interference: ☐ Narrowband ☐ WLAN ☐ ISM Equipment ☐ Other _____
Notes: _____

Point 5: _____
Values: Signal _____ Noise _____ SNR _____ Other _____
Interference: ☐ Narrowband ☐ WLAN ☐ ISM Equipment ☐ Other _____
Notes: _____

Point 6: _____
Values: Signal _____ Noise _____ SNR _____ Other _____
Interference: ☐ Narrowband ☐ WLAN ☐ ISM Equipment ☐ Other _____
Notes: _____

Point 7: _____
Values: Signal _____ Noise _____ SNR _____ Other _____
Interference: ☐ Narrowband ☐ WLAN ☐ ISM Equipment ☐ Other _____
Notes: _____

Point 8: _____
Values: Signal _____ Noise _____ SNR _____ Other _____
Interference: ☐ Narrowband ☐ WLAN ☐ ISM Equipment ☐ Other _____
Notes: _____

Point 9: _____
Values: Signal _____ Noise _____ SNR _____ Other _____
Interference: ☐ Narrowband ☐ WLAN ☐ ISM Equipment ☐ Other _____
Notes: _____

Point 10: _____
Values: Signal _____ Noise _____ SNR _____ Other _____
Interference: ☐ Narrowband ☐ WLAN ☐ ISM Equipment ☐ Other _____
Notes: _____

Obstacles in the immediate environment:

☐ Buildings   ☐ Trees   ☐ Body of water
☐ Hills   ☐ Towers   ☐ Warehouse shelves / goods
☐ Other _____
☐ Estimated Heights _____

Site Survey – BRIDGE Form

Environment where Bridge will be placed

☐ Outdoors on a building   ☐ Outdoors on a Mast   ☐ Outdoors on a Tower
☐ Retail Sales   ☐ Hallway / Corridor   ☐ Indoor Wiring Closet
☐ Manufacturing   ☐ Warehouse / Distribution
☐ Other _____

Elements to which the Bridge will be exposed

☐ Heat   ☐ Cold   ☐ Fluctuating Temperature
☐ Rain / Snow ☐ Dirt / Dust ☐ Grease
☐ Chemicals   ☐ Sunlight   ☐ Vibration
☐ Wind

Notes:
_____
_____
_____

Site Survey – Bridge Form   Certified Wireless Network Professional

Configuration Information

Bridge Management Information:

☐ HTTP _____

☐ Telnet _____

☐ SNMP _____

☐ Console/Serial Port _____

☐ Custom Application _____

Wireless VLANs:

ESSID _____ → VLAN _____   ESSID _____ → VLAN _____

ESSID _____ → VLAN _____   ESSID _____ → VLAN _____

ESSID _____ → VLAN _____   ESSID _____ → VLAN _____

ESSID _____ → VLAN _____   ESSID _____ → VLAN _____

ESSID _____ → VLAN _____   ESSID _____ → VLAN _____

ESSID _____ → VLAN _____   ESSID _____ → VLAN _____

ESSID _____ → VLAN _____   ESSID _____ → VLAN _____

ESSID _____ → VLAN _____   ESSID _____ → VLAN _____

IP Address: _____
MAC Address: _____
ESSID: _____

Authentication/Encryption:

☐ 802.1x/LEAP   ☐ 802.1x/PEAP   ☐ PPTP VPN   ☐ IPSec VPN
☐ L2TP VPN   ☐ KeyGuard   ☐ TKIP   ☐ MIC
☐ Other _____

Authentication Server Type:

☐ RADIUS   ☐ LDAP   ☐ Active Directory/Kerberos   ☐ eDirectory/NDS
☐ Other _____

Site Survey – Bridge Form   Certified Wireless Network Professional

Pictures of Bridge Mounting

In the following fields, record the name and description of each digital photograph taken.

Name: _____

Description: _____

Name: _____

Description: _____

Name: _____

Description: _____

Pictures of Antenna Mounting

In the following fields, record the name and description of each digital photograph taken.

Name: _____

Description: _____

Name: _____

Description: _____

Name: _____

Description: _____

Pictures of Custom Mounting Equipment (if required)

In the following fields, record the name and description of each digital photograph taken.

Name: _____

Description: _____

Name: _____

Description: _____

Name: _____

Description: _____

Site Survey – Workgroup Bridge Form   Certified Wireless Network Professional

Engineer Name: _____

Engineer Email: _____

Engineer Signature: _____

Customer Name: _____

Job Number: _____

WGB Name / Number: _____

WGB Location:

☐ Indoor   ☐ Outdoor

Building / Floor: _____

Floor Plan / Map Grid Reference: _____

AP to which WGB connects: _____

How many wired clients will be connected to the WGB? _____

Data Cabling

Cable path (from WGB location to Hub/Switch) _____

Cable Type / Length _____

WGB Type to be installed:

☐ 802.11
☐ 802.11a
☐ 802.11b
☐ 802.11g
☐ Other _____

Existing Network Connectivity Type:

☐ 10baseTx Hub   ☐ 10/100baseTx Switch
☐ 100baseTx Hub   ☐ 10/100baseFx Switch
☐ 10baseTx Switch   ☐ 10/100/1000baseTx Switch
☐ 100baseTx Switch

Ethernet Switch/Hub

Mfr. / Model _____

IP Address (if managed) _____

Site Survey – Workgroup Bridge Form Certified Wireless Network Professional

Name / Location _____

Port Number _____

Survey Data Rate:

☐ 1 Mbps ☐ 2 Mbps ☐ 5.5 Mbps ☐ 11 Mbps
☐ 6 Mbps ☐ 9 Mbps ☐ 12 Mbps ☐ 18 Mbps
☐ 24 Mbps ☐ 36 Mbps ☐ 48 Mbps ☐ 54 Mbps
☐ Proprietary _____

Total Throughput Required from WGB (if known) _____

Prevailing Traffic Types (for QoS purposes)

☐ FTP ☐ File Sharing ☐ Instant Messaging
☐ HTTP ☐ POP/SMTP ☐ Non-IP Protocols
☐ VoIP ☐ Data Backup ☐ Routing Protocols
☐ Video ☐ Database Access
☐ Telnet/SSH ☐ Warehouse Data

Types of Clients that will connect:

☐ Network Appliance ☐ Printer / Print Server
☐ Laptop PC ☐ Mobile Printers
☐ Desktop PC ☐ Game Station
☐ Set Top Boxes ☐ Scanner
☐ Other _____

Channel(s):

☐ 1 ☐ 2 ☐ 3 ☐ 4 ☐ 5 ☐ 6 ☐ 7 ☐ 8 ☐ 9 ☐ 10 ☐ 11
☐ 12 ☐ 13 ☐ 14 ☐ 36 ☐ 40 ☐ 44 ☐ 48 ☐ 52 ☐ 56 ☐ 60 ☐ 64

Output Power(s):

☐ 1 mW ☐ 2 mW ☐ 5 mW ☐ 10 mW ☐ 20 mW ☐ 32 mW
☐ 40 mW ☐ 50 mW ☐ 64 mW ☐ 100 mW ☐ 200 mW ☐ 500 mW
☐ 1 W ☐ Other _____

WGB Information

☐ WGB Type/Mfr. _____

☐ Antenna Type & Gain _____

☐ Pigtail Cable _____

☐ AC Power Cabling _____

Site Survey – Workgroup Bridge Form

☐ Surge Protection - ☐ YES ☐ NO

☐ PoE Injector (single, multiple) _____

☐ Lightning Protection (type, ohms) _____

☐ RF Cabling & Connectors (type, length, ohms) _____

Plenum Rating Required?

☐ YES ☐ NO

WGB Housing Type

☐ NEMA Enclosure
☐ Lockable Enclosure
☐ None
☐ Other _____

WGB Mounting Information

☐ Wall ☐ Mast
☐ Ceiling ☐ Tower
☐ Enclosure ☐ Roof
☐ Other: _____

☐ Ladder Required
☐ Lift Required
☐ Tower Climber Required

Mounting Height _____

Mounting Location _____

Orientation/Alignment _____

Mounting Gear Required _____

Identifying Landmarks / Items around WGB _____

Notes: _____

Antenna Mounting Information

☐ Wall ☐ Mast
☐ Ceiling ☐ Tower
☐ Enclosure ☐ Roof
☐ Other: _____

Site Survey – Workgroup Bridge Form					Certified Wireless Network Professional

☐ Ladder Required
☐ Lift Required
☐ Tower Climber Required

Mounting Height _____

Mounting Location _____

Polarization: ☐ Vertical ☐ Horizontal ☐ Circular

Orientation/Alignment _____

Mounting Gear Required _____

Notes: _____

Measurement Points:

Point 1: _____
Values: Signal _____ Noise _____ SNR _____ Other _____
Interference: ☐ Narrowband ☐ WLAN ☐ ISM Equipment   ☐ Other _____
Notes: _____

Point 2: _____
Values: Signal _____ Noise _____ SNR _____ Other _____
Interference: ☐ Narrowband ☐ WLAN ☐ ISM Equipment   ☐ Other _____
Notes: _____

Point 3: _____
Values: Signal _____ Noise _____ SNR _____ Other _____
Interference: ☐ Narrowband ☐ WLAN ☐ ISM Equipment   ☐ Other _____
Notes: _____

Point 4: _____
Values: Signal _____ Noise _____ SNR _____ Other _____
Interference: ☐ Narrowband ☐ WLAN ☐ ISM Equipment   ☐ Other _____
Notes: _____

Point 5: _____
Values: Signal _____ Noise _____ SNR _____ Other _____
Interference: ☐ Narrowband ☐ WLAN ☐ ISM Equipment   ☐ Other _____
Notes: _____

Point 6: _____
Values: Signal _____ Noise _____ SNR _____ Other _____
Interference: ☐ Narrowband ☐ WLAN ☐ ISM Equipment   ☐ Other _____
Notes: _____

Point 7: _____

Values: Signal _____ Noise _____ SNR _____ Other _____
Interference: ☐ Narrowband ☐ WLAN ☐ ISM Equipment ☐ Other _____
Notes: _____

Point 8: _____
Values: Signal _____ Noise _____ SNR _____ Other _____
Interference: ☐ Narrowband ☐ WLAN ☐ ISM Equipment ☐ Other _____
Notes: _____

Point 9: _____
Values: Signal _____ Noise _____ SNR _____ Other _____
Interference: ☐ Narrowband ☐ WLAN ☐ ISM Equipment ☐ Other _____
Notes: _____

Point 10: _____
Values: Signal _____ Noise _____ SNR _____ Other _____
Interference: ☐ Narrowband ☐ WLAN ☐ ISM Equipment ☐ Other _____
Notes: _____

Known Dead Spots:

Point 1: _____
Caused by: _____
Suggestions: _____

Point 2: _____
Caused by: _____
Suggestions: _____

Point 3: _____
Caused by: _____
Suggestions: _____

Site Survey – Workgroup Bridge Form

Obstacles in the immediate environment:

☐ Metal blinds ☐ Fire doors ☐ Metal mesh windows
☐ HVAC ☐ Duct Work ☐ Firewall
☐ Elevator ☐ Machinery ☐ Warehouse shelves / goods
☐ Pipes
☐ Other _____

Environment where WGB will be placed

☐ Open office space ☐ Office with cubicles ☐ Warehouse / Distribution
☐ Retail Sales ☐ Freezer / Cold Storage ☐ Hallway / Corridor
☐ Manufacturing ☐ Outdoors
☐ Other _____

Elements to which the WGB will be exposed

☐ Heat ☐ Cold ☐ Fluctuating Temperature
☐ Rain / Snow ☐ Dirt / Dust ☐ Grease
☐ Chemicals ☐ Sunlight ☐ Vibration
☐ Wind

Notes:
_____
_____
_____

Configuration Information

WGB Management Information:

☐ HTTP _____

☐ Telnet _____

☐ SNMP _____

☐ Console/Serial Port _____

☐ Custom Application _____

IP Address: _____
MAC Address: _____
ESSID: _____

Authentication/Encryption:

☐ 802.1x/LEAP   ☐ 802.1x/PEAP   ☐ PPTP VPN   ☐ IPSec VPN
☐ L2TP VPN   ☐ KeyGuard   ☐ TKIP   ☐ MIC
☐ Other _____

Authentication Server Type:

☐ RADIUS   ☐ LDAP   ☐ Active Directory/Kerberos   ☐ eDirectory/NDS
☐ Other _____

Pictures of Workgroup Bridge Mounting

In the following fields, record the name and description of each digital photograph taken.

Name: _____

Description: _____

Name: _____

Description: _____

Name: _____

Description: _____

Pictures of Antenna Mounting

In the following fields, record the name and description of each digital photograph taken.

Name: _____

Description: _____

Name: _____

Description: _____

Name: _____

Description: _____

Pictures of Custom Mounting Equipment (if required)

In the following fields, record the name and description of each digital photograph taken.

Name: _____

Description: _____

Name: _____

Description: _____

Name: _____

Description: _____

Performing the Site Survey with AirMagnet

December 1, 2002

Background

A thorough Site Survey is the linchpin in the successful deployment of any wireless network. The rise of 802.11 technology has brought a much needed upgrade to the overall flexibility of the modern network, and in the process, spurred an increase in end-user productivity. These benefits, however, have also been accompanied by a susceptibility to environmental factors, which can degrade network performance. It is the purpose of the Site Survey to mitigate these factors and insure that the actual performance observed in the network meets the needs of all the network's users. In practice, this translates to making sure each station in the network receives strong signals and exhibits high transmission throughput rates in its area of operation. This paper is written with these goals in mind, and details the process of performing a wireless Site Survey with the AirMagnet Handheld Wireless Analyzer.

Prior to Performing the Site Survey

Each wireless project is unique, and a clear understanding of the demands placed on the network is essential before beginning the site survey. While the site survey will uncover potential problem areas in the network, it is crucial to have a grasp of basic usage information such as the number of concurrent users the network will need to support, and any areas where user concentration will be particularly high.

In addition to getting a handle on the demands that will be placed on the network, installers will need to take into account several limiting factors inherent to 802.11 networks.

DISTANCE

One of the most obvious considerations when planning a wireless network is the distance between the APs and the stations they will support. In theory, signal strength and distance will exhibit an inverse square relationship according to the formula:

$$s = \frac{P_t}{4\pi d^2}$$

This formula is based on an ideal antenna where **s** is signal strength, **Pt** is transmission power, and **d** is distance. Most commercial wireless transmitters and receivers attempt to combat this trend using a variety of technologies, but the overarching trend remains - as distance increases, signal quality will decrease leading to an adverse effect on performance.

ENVIRONMENT

In addition to distance, the network's environment will have a strong impact on the observed performance in different locations. Obstructions between the AP and stations lead to defraction, refraction, and reflection of the RF signal, which in turn leads to a corresponding degradation in signal quality. However, these effects can be minimized through the use of a few straight-forward practices:

Antenna Placement

Minimizing the obstructions in the Line of Sight between the AP and station will help minimize signal degradation. In an office environment with cubicles, placing the antenna in a higher location will generally improve signal quality to the various stations. In environments using multiple antennas, each antenna should be placed at least one wavelength apart. However, when operating in the 2.4 GHz range, this can generally be accomplished by making sure antennas are placed at least six inches apart (antennas operating in the 5 GHz band are even less susceptible to this issue).

Identifying Metal Obstructions

RF signals are far more susceptible to degradation by metal obstructions than other materials. Consequently, care should be taken when placing large metal objects such as storage cabinets and shelving in the environment.

© AirMagnet Inc. 2003   www.airmagnet.com

Building Materials

Given the effect that metals can have on signal quality, it is important to note the location of steel and other metal materials in the site infrastructure. The building's blueprints can usually assist in this regard, and can help to identify potential problem areas where readings should be taken during the site survey.

Sources of Interference

An array of devices operate in or near the 2.4 GHz, and can be a significant source of noise in 802.11b networks. These devices include cordless phones, microwave ovens, as well as other common wireless technologies such as Bluetooth devices. These signals can cause nodes to receive "damaged" 802.11 frames, or can cause to the node to wait indefinitely for the channel to clear. If possible, the site survey should be performed when these devices are operating normally in order to provide a realistic representation of noise in the environment.

COMMON SOURCES OF INTERFERENCE

 2.4 GHz Cordless Phones

 Microwave Ovens

 BlueTooth Devices

 Other Wireless Networks

 Wireless Cameras

Performing the Site Survey

The AirMagnet family of wireless analyzers provide an intuitive and efficient means of performing a comprehensive site survey. AirMagnet Analyzers can collect information based on event, time interval, or user defined location. This provides users the flexibility to capture the information that is most relevant to the task, while limiting extraneous data.

Before beginning, it is a good practice to plan and document the path the user will follow during the Site Survey. Remember, with AirMagnet, the Site Survey can be paused and resumed at will, so the entire survey need not be completed in one continuous walk-through. Additionally, specific points of interest should be identified either from a physical inspection of the site or from reviewing building blueprints or site maps. These points could be areas where user concentration will be particularly high or simply waypoints on the path of the survey.

With the plan complete, the user is free to begin the Site Survey. The following section details the process of performing the Site Survey with an AirMagnet Wireless Analyzer. Note that the screens in the following sections were taken from The AirMagnet Handheld, but the process is applicable across all AirMagnet products.

The Site Survey Process

1. Open AirMagnet on your Pocket PC.

2. Select the Tools icon at the bottom of the screen.

3. From the Tools Page select "Survey" (1). It may be necessary to scroll to the left if the Survey Tab is not initially visible.

© AirMagnet Inc. 2003   www.airmagnet.com

MAKING SETTINGS FOR THE SITE SURVEY

4. Select "SSID" from the AP/SSID drop down menu at the top-left of the screen. ② Selecting SSID will allow you to monitor multiple APs during your survey. Conversely, select "AP" only if you wish to survey one particular AP.

5. Choose the appropriate SSID from the adjacent drop down list.

6. Tap the Configuration icon 🔲, to go to the Survey Configuration screen shown in Figure 2.

7. Input a name for the survey in the "Log file" field. ③

8. Check/Uncheck "Log by Event" - The Log by Event feature ④, allows for a very detailed analysis of the environment, based on the following 3 types of events:

    Associated with AP - Takes a data point when there is a change of AP that the AirMagnet is associated with. This indicates the boundary between competing APs. Network users who are located on a boundary point, regularly see a degradation in performance due to the hardware switching between the two APs

    Signal Strength Change - Notes all changes in signal strength. Normally, this feature will generate a large number of data points, and as a result should only be used when obtaining a detailed scan of a particular area.

    Speed Change - Records changes in transmission speed between the AP and the AirMagnet. These data points are very significant when planning station placement in the network. Stations operating at sub-optimal speeds will generate chronic user complaints, and can bog down the entire network.

9. Check/Uncheck "Log by Time" - This option allows data to be taken at regular time intervals as defined by the user. The user can select the time interval ⑤, which ranges from once every second to once every 3 minutes.

10. When Configurations are complete tap the 'OK' button at the bottom of the screen.

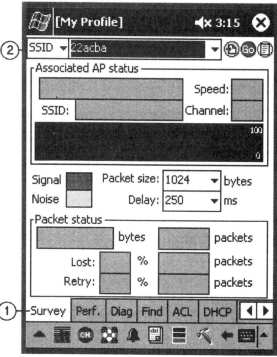

Figure 1 - Survey Screen

Figure 2 - Configuration Screen

© AirMagnet Inc. 2003   www.airmagnet.com

TAKING DATA

1. Go to Location 1 according to your survey plan.

   Tap the 🔵 icon in the top right hand corner of the Site Survey screen. You should begin seeing real time RF and network performance data in the survey screen ①.

   Note: If you receive a "No WEP Configuration" message, see *Configuring WEP on the AirMagnet*. at the end of this document.

2. With the Survey running, tap the Configuration icon 🔧. "Location 1" should automatically be entered in the Location Field. While standing at Location 1, tap the "OK" button at the bottom of the page. This records the current data and names it according to the entry in the Location Field. Repeat this process as needed to take manual readings of other important locations.

3. Walk the prearranged path to Location 2 at a controlled steady pace. AirMagnet will automatically collect data according to selections made in the Log by Event and Log by Time sections mentioned earlier. If selected, AirMagnet will beep to indicate when a data point has been taken.

4. Take additional data points at Location 2...n as described in step 2.

5. When the Site Survey is complete, press the Stop icon ✖, shown in Figure 3.

ANALYZING THE COLLECTED DATA

All data captured during the Site Survey is saved in a .csv file for convenient analysis. All captured data can be easily charted using basic functionality in Microsoft Excel as shown in Figure 5.

The Site Survey file can be moved from your Pocket PC to your Desktop by using the ActiveSync software that came with the your handheld. The file can be found choosing the Explore button ① on the top panel of the ActiveSync window, and selecting the AirMagnet folder within.

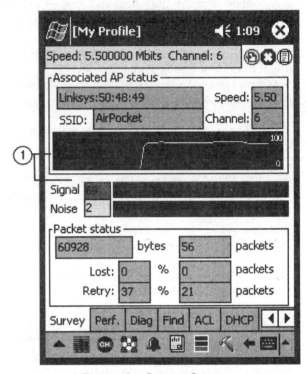

Figure 3 - Survey Screen

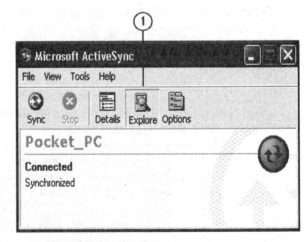

Figure 4 - Active Sync

Once the file is on your PC, it can be opened using Microsoft Excel. The following section gives an explanation of each data field included in the file, and their significance.

Loc./Event - This field identifies the data point as either a specific Event, a Log By Time, or a specified Location depending on the configuration used for the site survey.

Time - Shows the Time that the data point was taken.

AP - Provides the name of the AP for which data has been taken. This name is defined on the AP configuration.

MAC - Provides the MAC Address for the AP

Signal (%) - Provides a measure of Signal Strength in a range from 0 to 100%.

Noise (%) - Provides a measure of Noise Level in a range from 0 to 100%. As mentioned in earlier sections, noise can arise from any number of sources. This reading reflects true noise in the spectrum and is distinct from the natural "bleedover" associated with an active 802.11 channel.

Signal (dBm) - Provides a measure of signal strength in a range of -99 to -10 dBm. (The unit dBm represents power relative to one milliwatt and is represented by the formula Pdbm = 10logPmW).

Noise (dBm) - Provides a measure of Noise Level in a range from -99 to -10 dBm.

Speed - Displays the Communication Speed between AirMagnet and the AP in Mbits/sec. The 802.11 spec supports speeds of 1,2,5, and 11Mbits/sec. As Signal Quality decreases, so will the maximum connection speed.

Channel - Designates the Channel on which the AP is operating.

Packet Loss - Gives the percentage of packets sent between AirMagnet and the AP, which have been lost.

Packet Retry - Gives the percentage of packets between AirMagnet and the AP, which were retried. This can be a telling statistic when evaluating the Site Survey. If a station receives packets with errors, it simply does not Ack back to the Sender, and waits for the packet to be resent. These errors can be caused by a variety of sources including collisions between transmitting stations or interference in the spectrum. A significant rate of packet retries may warrant further investigation to determine the source of the problem.

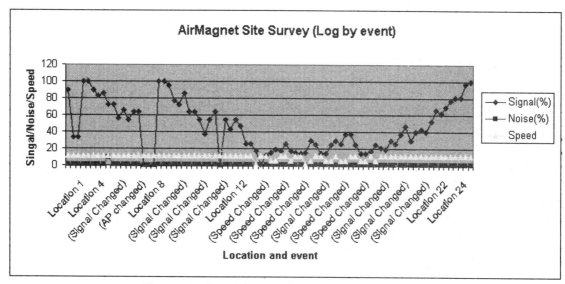

Figure 5 - Excel Graph of Survey Data

Performance Guidelines

When evaluating the Site Survey, it is recommended that users pay close attention to Speed, Signal-Noise Ratio, Packet Loss, and Packet Retry data, as each of these statistics are crucial to predicting service coverage. The following guidelines provide a rule of thumb when evaluating these areas of the site survey.

Signal/Noise - The signal-to-noise ratio should always be greater than 30%.

Speed - In most environments, speed should be kept greater than or equal to 5.5 Mbps.

Packet Loss and Packet Retry - Both of these levels should remain below 3%.

RSSI, dBm, and %

Signal levels are regularly reported in either RSSI (Received Signal Strength Indication), dBm, or Percentage. This has lead to quite a bit of confusion concerning how measurements taken in one set of units relates to measurements taken in another. To help mitigate this issue, we have included a conversion chart (left) to provide a quick way of mapping between these three units.

RSSI	%	dBm	RSSI	%	dBm
1	1	-108	51	72	-60
2	1	-107	52	73	-59
3	1	-106	53	75	-58
4	1	-105	54	77	-57
5	1	-104	55	78	-56
6	1	-103	56	80	-55
7	1	-102	57	82	-55
8	1	-101	58	83	-54
9	1	-100	59	85	-53
10	1	-100	60	87	-52
11	1	-99	61	88	-51
12	1	-98	62	90	-50
13	1	-97	63	93	-49
14	1	-96	64	95	-48
15	1	-95	65	98	-47
16	1	-94	66	100	-45
17	1	-93	67	100	-44
18	1	-92	68	100	-43
19	1	-91	69	100	-42
20	5	-90	70	100	-40
21	10	-90	71	100	-39
22	12	-90	72	100	-38
23	14	-89	73	100	-37
24	15	-88	74	100	-36
25	17	-87	75	100	-35
26	19	-86	76	100	-34
27	21	-85	77	100	-33
28	23	-84	78	100	-32
29	25	-83	79	100	-31
30	26	-82	80	100	-30
31	28	-81	81	100	-30
32	30	-80	82	100	-29
33	33	-79	83	100	-28
34	35	-78	84	100	-27
35	38	-77	85	100	-26
36	40	-75	86	100	-25
37	43	-74	87	100	-24
38	45	-73	88	100	-23
39	48	-72	89	100	-22
40	50	-70	90	100	-21
41	52	-69	91	100	-20
42	54	-68	92	100	-19
43	56	-67	93	100	-17
44	58	-66	94	100	-15
45	60	-65	95	100	-14
46	62	-64	96	100	-12
47	64	-63	97	100	-10
48	66	-62	98	100	-10
49	68	-61	99	100	-10
50	70	-60	100	100	-10

Accounting for Variance in Wireless Cards

Wireless network cards manufactured by different vendors can show different results based a variety of variables in the cards themselves, such as the card's transmission power, differing uses of AGC (Automatic Gain Control), and a host of other factors. While these differences do not expressly mean one card is superior to another, the network engineer should be aware of them and understand how readings taken during the site survey relate to the performance users can expect when using a given card.

To assist in this regard, AirMagnet performed a study across multiple card vendors to give users a point of reference when estimating performance between cards. The study was performed with the AirMagnet Handheld using compatible cards from Cisco, Proxim, and Symbol. The study was performed in a traditional office environment with a single exterior wall between the Access Point and the AirMagnet.

The results of these tests are shown in Figure 6 and Figure 7. These results should only serve as a guideline when estimating performance across different cards, and similar tests should be performed in your local environment to quantitatively adjust for readings in a given network.

Signal Level (dBm)

		1	2	3	4	5	6	7	8	9	10	11	12	13	14	15
	distance	10'	20'	30'	40'	50'	60'	70'	80'	90'	100'	110'	120'	130'	140'	150'
AirMagnet w/Symbol		-28	-36	-38	-46	-50	-53	-60	-55	-60	-68	-73	-73	-75	-80	-79
AirMagnet w/Cisco		-48	-58	-62	-73	-72	-73	-75	-79	-80	-82	-82	-82	-82	-85	-87
AirMagnet w/Proxim		-21	-20	-40	-51	-50	-73	-76	-72	-81	-86	-88	-75	-97	-98	-99

Network Speed (Mbps)

		1	2	3	4	5	6	7	8	9	10	11	12	13	14	15
	distance	10'	20'	30'	40'	50'	60'	70'	80'	90'	100'	110'	120'	130'	140'	150'
AirMagnet w/Symbol		11	11	11	11	11	11	11	11	11	11	11	11	11	5.5	11
AirMagnet w/Cisco		11	11	11	11	11	11	11	11	11	11	11	11	11	5.5	11
AirMagnet w/Proxim		11	11	11	11	5.5	5.5	2	5.5	5.5	2	11	1	1	1

Figure 6 - Observed Variances in Wireless Network Cards

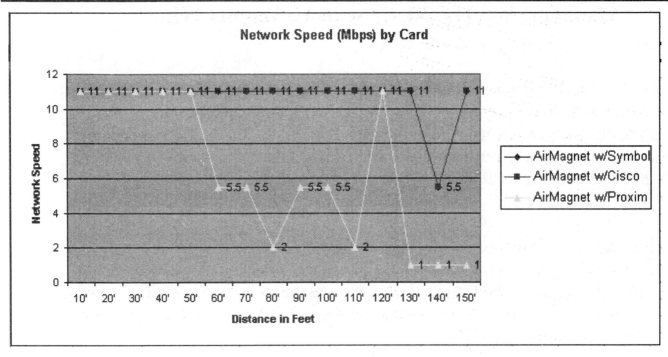

Figure 7 - Observed Variances in Wireless Network Cards

Configuring WEP on the AirMagnet Handheld

1. In order to perform a site survey, you may be required to configure WEP on the AirMagnet Handheld to work with the SSID/APs in your network. This is a very straightforward process and is detailed below.

2. Open AirMagnet

3. Select the (up arrow) symbol in the bottom left corner of the screen, and choose Configure... from the pop-up menu.

4. Select the 802.11 tab from the bottom of the Configuration Screen.

5. Select the appropriate SSID from the SSID pull-down menu.

6. Tap the WEP button

7. Select Char or Hex depending on the key in use.

8. Select Key Length

9. Select Field 1 and enter the appropriate key for the SSID

10. Tap Apply.

Managing 802.11g WLAN with AirMagnet Trio

With the 802.11g standard approved by the Institute of Electrical and Electronic Engineers (IEEE), it is generating a great deal of interest among wireless users. It can be compared as second only to the interest generated during the introduction of the 802.11b standard. This standard makes available high data rates comparable to the 802.11a standard, but most importantly provides backward compatibility to the widely implemented 802.11b standard.

This white paper explains the basics of the 802.11g standard and the deployment caveats with specific information on the importance of obtaining detailed 802.11g traffic information on the WLAN. It also explains the co-existence of 802.11b and g devices in the same WLAN environment and how 802.11g WLANs are managed with the AirMagnet analyzers.

Introduction to 802.11g

Just like 802.11b devices, 802.11g devices operate in the 2.4 Ghz Industrial Scientific Medical (ISM) band except it uses the Orthogonal Frequency Division Multiplexing (OFDM) technology, unlike the Complementary Code Keying (CCK) modulation used by the 802.11b standard. OFDM is also used by 802.11a devices that operate in the 5Ghz Unlicensed National Information Infrastructure (UNII) band. The 802.11g standard also supports Barker Code and Complementary Code Keying (CCK) modulation giving 1, 2, 5.5 and 11 Mbps data rates for backward compatibility with the 802.11b standard. OFDM provides 6, 9, 12, 18, 24, 36, 48 and 54 Mbps data rates. The optional Packet Binary Convolution Coding (PBCC) encoding method provides data rates of 22 and 33 Mbps. The standard only includes data rates 1, 2, 5.5, 11, 6, 12 and 24 as being mandatory for transmission and reception. Similar to the 802.11b standard, 802.11g devices are limited to three non-overlapping channels and the new physical layer is called the Extended Rate Physical (ERP) layer. Traditionally, though 802.11b and 802.11g devices communicate using CCK and OFDM modulation schemes respectively and because of backward compatibility 802.11g devices have to support both modulation schemes.

Every 802.11 packet is made of the preamble, header information and the payload. The preamble announces the beginning of the wireless transmission and influences the network throughput with the longer preamble lowering the throughput of the network compared to the shorter preamble. The use of the short preamble (96 μsec) is optional in the 802.11b standard, while the longer preamble (120 μsec) is mandatory. 802.11g supports both short and long preambles. 802.11g employs Carrier Sense Multiple Access/ Collision Avoidance (CSMA/CA). The CSMA/CA protocol allows a device that is transmitting data exclusive transmitting rights. No other 802.11 device will transmit at that time and will begin transmission only when the medium is clear. This avoids collisions after waiting for random intervals of time. This random back-off interval is calculated using slot times multiplied with a random number.

Working in Harmony

The slot time is different for 802.11b and g devices. This slot time not only lowers the throughput in a pure 802.11b or g environment, it also reduces the throughput to undesirable levels in a mixed 802.11b and g environments. The 802.11g devices (9 μsec) now have to use the longer slot time of 802.11b devices (20 μsec). This is done to avoid preferential treatment for devices with shorter slot times transmitting first all the time. This makes it very important for a wireless administrator to be aware of the nature of devices in the WLAN.

OFDM signals are also not heard by the 802.11b devices which will incorrectly assess the medium to be free for transmission leading to collisions and reducing throughput. The 802.11g standard (section 7.3.2.13 ERP information element) requires devices to employ protection mechanisms to improve the performance in a mixed 802.11b/g environment. The 802.11g standard allows the use of two protection mechanisms in such an environment: *RTS/CTS and CTS-to-self*. These protection mechanisms prepare 802.11g devices for communication with a 802.11b devices using the CCK modulation scheme. In the *RTS/CTS protection mechanism*, the device wanting to communicate sends a Request-To-Send (RTS) message to the destination node. It then receives a Clear-To-Send (CTS) message from the destination node indicating that the RTS message was received and that the data packet can be sent. This CTS message is heard by all the devices on the network and now know that some transmission is taking place and will cease their own transmission appropriately. After receiving the CTS signal the device sends data and waits for an ACK signal to verify that a successful transmission has transpired without collisions. In the *CTS-to-self protection mechanism* the 802.11g AP will send a CTS message when it desires to send data, even though there was no RTS message received. Both these mechanisms help reduce collisions.

Unhealthy WLANs

The immense benefit of backward compatibility is overshadowed by the lowering of performance of the wireless network in a mixed environment. A WLAN with 802.11g only devices will provide higher throughput than what it will in a mixed environment with 802.11b devices. 802.11a devices do not impact the performance as they operate in the separate 5Ghz spectrum. Different configurations on the 802.11g access points and 802.11b and g client devices will cause a variety of problems on the WLAN. Problems could range from minor ones like devices sending frames at low speeds and use of non-standard speed transmission rates; to major problems such as having a few client devices not using mechanisms to operate efficiently in a mixed environment. Also in certain cases these mechanisms, which are discussed later in the note, may cause more problems than what it attempts to solve. This makes it very important to understand minor details of various factors during implementation and maintaining 802.11g networks that will ultimately govern the solution which is unique for every 802.11g WLAN.

AirMagnet - Keeping a Watchful Eye

The AirMagnet analyzers help network professionals administer and troubleshoot WLANs, eliminate connection problems, maintain network performance levels and maintain a high level of network security. The AirMagnet Analyzer's product suite provides simultaneous viewing of all the channels with the ability to drill down to any detail on any node in the network. The analyzer monitors the WLAN environment for a wide range of conditions that can cause performance problems and alerts the user when such a condition occurs. The analyzers allows the user to select and view up to six real-time graphs to monitor the status of the network.

The pie chart on the **Start** page of the AirMagnet wireless analyzers will indicate the standards followed by the devices on the network. The following figure shows the division of the number of Access points (AP) and stations (STA) following different standards.

Detecting the presence of 802.11b devices on the WLAN will make the administrator aware of the likelihood of performance problems occurring on the WLAN.

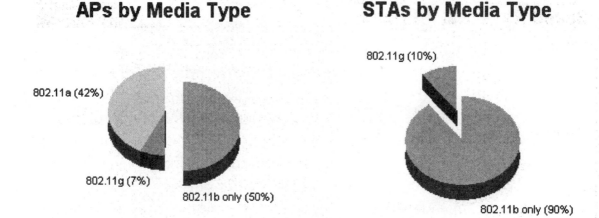

Figure 1: Access points (AP) and stations (STA) by 802.11 standards.

The **Channel** page also announces the presence of any b-only APs or stations operating in a particular channel. The Figure below shows two b-only AP out of the three operating in channel 1 and one b-only station out of two on channel 1.

Channel Detail	
Last updated time	10:14:35.080000
Channel	1 (2412 MHz)
# AP	3 (2 b Only)
# STA	2 (1 b Only)
Scan time	250 ms

*Figure 2: **Channel** page reports a mixed 802.11b/g environments.*

Need for Speed

APs use beacons and stations in association requests and responses to exchange supported data rates. An 802.11g client can transmit data frames up to a maximum of 54Mbps. The device will transmit at the lower standardized rates if the signal strength and link quality decrease due increasing distance between devices or the presence of interfering devices, such as microwave ovens, 2.4 Ghz cordless phones or neighboring inband radios on the same frequency. This is called Adaptive Rate Selection (ARS) or Dynamic Rate Shifting (DRS). The speed at which the frames are being transmitted and received are a good indication of the overall performance of the network and also help in planning cell sizing and security. The **Channel** and **Infrastructure** pages report data rates for the transmitted and received frames.

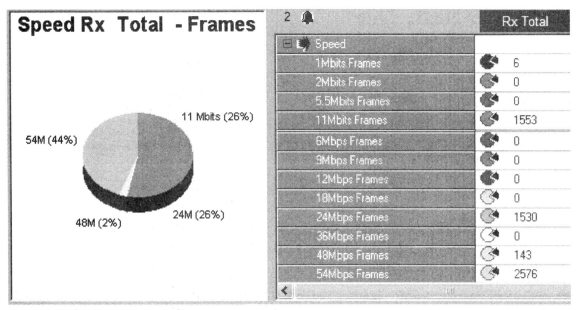

*Figure 3: Data rates for a receiving AP or Station on the **Infrastructure** page.*

The **Channel** and **Infrastructure** page seperate 802.11b and g traffic in terms of frames and bytes for channels, APs and Stations.

Speed		
Media Type		
802.11b Frames	4443	36 %
802.11g Frames	7856	63 %
802.11b Bytes	155656	3 %
802.11g Bytes	4121980	96 %
Alert	0	
Frames/Bytes	12299	4277636
Ctrl. Frames/Bytes	8396	285464
Mgmt. Frames/Bytes	69	6724
Data Frames/Bytes	3159	3525392
Channel Detail		

Figure 4: 802.11b/g traffic for channels, APs and STAs.

The overall utilization and throughput division by the 802.11b and g devices is provided on the **Channel** page of the AirMagnet handheld and laptop analyzer.

a) Utilization and throughput for a channel on the basis of media-type: In the figure below when filtered by media type, 49.59% of the total utilization (39.84%) on channel 5 is due to 802.11b traffic and 50.41% due to 802.11g traffic. Also 802.11b traffic accounts for only 14.56% of the throughput on that channel and 85.44% is caused due to 802.11g traffic.

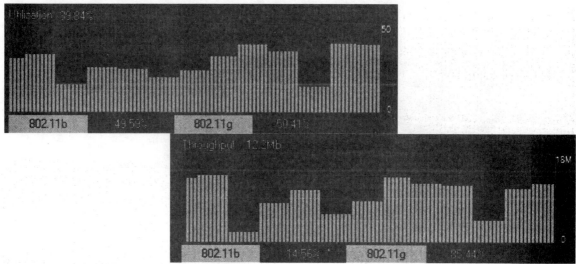

Figure 5: Utilization and throughput for 802.11b/g devices on the **Channel** *page.*

In the figure below, utilization due to 802.11b traffic for the channel is 13% and 22% for 802.11g traffic.

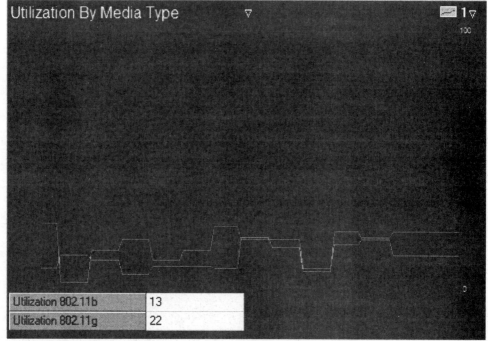

.Figure 6: Utilization on a channel due to 802.11b/g traffic on the **Channel** *page.*

b) Utilization and throughput for a channel on the basis of speed: This graph on the **Channel** page indicates the overall utilization of the channel due to individual data rates. For example, in the figure below, 12% of the overall utilization (36.41%) on channel 5 can be accounted to 1Mbps data rate, 24% to 11Mbps, 3% to 24Mbps, 58% to 54Mbps frames. Also 7% of the overall throughput on that channel is due to 11Mbps frames, 2% by 24Mbps and 89% by 54Mbps frames.

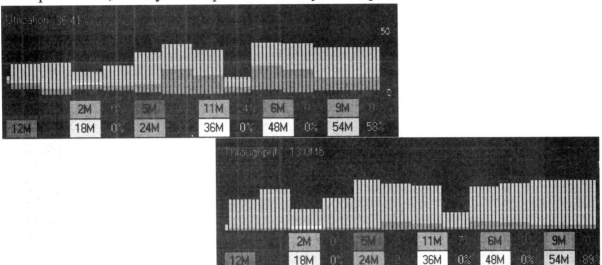

*Figure 7: Percent utilization and throughput by speeds on the **Channel** page.*

The figure below provides in Kbits/sec the rate at which 802.11b and g devices are communicating.

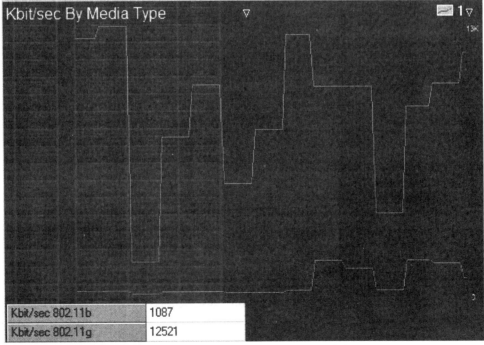

*Figure 8: Kbit/sec for the two standards on the **Channel** page.*

©2002-2003 AirMagnet Inc, All rights reserved.

AirMagnet, AirWISE, PocketNOC, the AirMagnet logo are trademarks of AirMagnet Inc.

All other product names mentioned herein may be trademarks of their respective companies.

The **Charts** screen displays top traffic generating APs, stations, and channels on the WLAN. Selecting the media-type option from the drop down menu provides information on the basis of 802.11a, b and g traffic. This screen provides a one-stop view for the top talkers on the WLAN. The figure below shows a few of the top ten APs in the WLAN network. Different color codes denote the 802.11a, b or g traffic to/from the AP. Clicking on individual APs will take you to the Infrastructure page to get more detailed information on that particular AP.

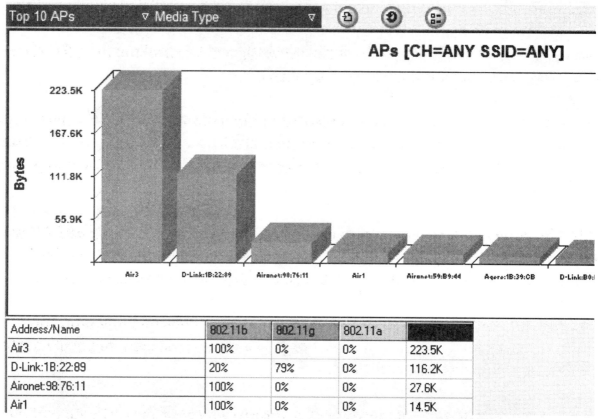

*Figure 9: Top 10 APs based on media-type displayed on the **Charts** screen.*

AirWISE - The Guru of Performance Alerts

AirMagnet Wireless System Expert (AirWISE) is the wireless performance and security monitoring platform that collects network performance statistics and reports various performance issues occurring in the 802.11g-only and mixed mode environment. These problems may severely affect the throughput of the network and need to be addressed immediately. **AirWISE** runs continuously in the background and automatically collects WLAN performance statistics and not only generates alarms for problems, but also provides solutions to handle them. Listed below are some of the performance alarms generated by **AirWISE**.

Low speed tx rate exceeded: There may be devices in the WLAN that may be transmitting most of the frames at lower speeds which lowers the throughput of the WLAN. This can be avoided if better signal strength or quality is provided.

802.11g AP using short time slot: As discussed earlier, an 802.11b STA communicating to a 802.11g AP forces the 802.11g AP to use a longer time slot. Incorrectly used time slots by the AP in a mixed environment causes collisions impacting the

throughput. This condition can be further investigated by checking the CRC errors and frame retries on the channel used by the AP.

802.11g pre-standard devices implemented in the network: **AirWISE** detects pre-standard devices not employing the protection mechanisms that need to have their firmware upgraded to be compliant to the approved 802.11g standard and improve throughput.

802.11g stations not implementing protection mechanism in a mixed 802.11b/g environment: 802.11g clients not using the protection environment in a mixed 802.11b/g environment need to be detected and dealt with to reduce collisions.

802.11g AP implementing wrong protection: APs not implementing protection mechanisms or implementing incorrect ones in a mixed mode environment should be detected to help reduce collisions.

Implementation of non-standard speed transmission rates: Few APs support non-standard transmission speed rates such as Super G, turbo mode, Packet Burst, etc. These vendor-proprietary higher speed rates increase chances of interference with devices in the neighboring channels. During implementation of devices with proprietary rates, it is important to plan the channel allocation for the respective devices and test them in a multi-vendor environment.

802.11g protection overhead: The RTS and CTS messages use the slower CCK modulation at 802.11b speeds. Implementing protection mechanisms such as RTS/CTS and CTS-to-self increases time for data transfer and processing which invariably reduces effective throughput.

The use of protection mechanisms when 802.11b devices generate only a small fraction of the overall WLAN traffic in a mixed 802.11b/g environment is **not** recommended as it generates unwanted overhead, it is **strongly advised to enable** these mechanisms as the number of 802.11b clients and their respective traffic increases.

> [10/10 11:08:20 - 10/10 11:08:20]
> 802.11g AP D-Link:99:5D:FE (SSID:unknown) is currently beaconing for 802.11g protection mechanism for 802.11g/11b mixed mode operation in channel 5. IEEE 802.11g standard specifies protection mechanism to keep 802.11b devices from interfering with 802.11g devices at the cost of lower performance than a pure 802.11g deployment without protection mechanism. Since an 802.11b device (using CCK modulation) cannot detect 802.11g signal (using OFDM modulation) in the same 2.4GHz spectrum, the 802.11b device may transmit over an 802.11g OFDM transmission causing packet collisions. IEEE 802.11g standard protection mechanism is implemented by 802.11g devices using 802.11b RTS/CTS or CTS-to-self control frames to coordinate packet transmission and clear channel assessment (CCA) among 802.11g and 802.11b devices.
>
> Even though RTS/CTS and CTS-to-self help minimizing packet collisions, they do add overhead to each packet transmission. In the case where your pure 802.11b devices constitute only a very small fraction of your WLAN traffic, it may be more efficient to configure your WLAN to not use the protection mechanism. Your AP configuration may allow you to turn on or off such protection mechanism. In order to determine the optimum configuration for protection mechanism (on or off), please use AirMagnet channel view to profile your 802.11g and 802.11b traffic load. While using AirMagnet channel view, please note packets of speed 1,2,5.5, and 11mbps are pure 802.11b (using CCK modulation) traffic where other speeds are 802.11g (using OFDM modulation) traffic.
>
> Another way to optimize for the maximum 802.11g throughput is to migrate all of your 802.11b devices to 802.11g. The goal would be a pure 802.11g WLAN thus 802.11g protection mechanism will not be activated by your APs. Some of the 802.11b devices to be migrated to 802.11g in this channel are:
>
> AP: Aironet:59:B9:44 (SSID:funk)

*Figure 10: Detailed explanation for the 802.11g protection overhead alarm on the **AirWISE** screen.*

Device threshing between 802.11g and 802.11b: Dual mode(802.11b/g) clients in the network may switch between OFDM and CCK technologies transmitting at different speeds. Such devices need to be detected and configured correctly to avoid this problem.

Summary

With high data rates and backwards compatibility for 802.11b devices, the 802.11g standard has a promising future in Enterprise and SOHO environments. But determining and understanding the fine boundary between 802.11b and g devices is a must before any implementation. Effective administration of a 802.11g WLAN requires network professionals to maintain complete control over the access points operating in the environment and their configuration. Additionally, misconfiguration and movement of 802.11b and g devices have a significant impact on network performance and reliability. AirMagnet's Handheld Analyzer, Laptop Analyzer and Distributed System provide intricate details required to plan, maintain and troubleshoot 802.11g networks. These products organize information in a logical hierarchical structure to give the user information that is most useful with the control and flexibility to access more detailed information as they need it.

Converting Signal Strength Percentage to dBm Values

Joe Bardwell, VP of Professional Services

Executive Summary

WildPackets' 802.11 wireless LAN packet analyzers, AiroPeek and AiroPeek NX, provide a measurement of RF signal strength represented by a percentage value. The question sometimes arises as to why a percentage metric is used, and how this relates to the actual RF energy that is present in the environment. This paper discusses RF technology with sufficient detail to provide a basis for understanding the issues related to signal strength measurement.

November 2002

Contents

Measurement Units for RF Signal Strength 1

The mW and dBm Units of Measure 1

The Receive Signal Strength Indicator (RSSI) 3

RSSI in the 802.11 Standard 4

Granularity in RSSI Measurements 4

The Choice of a Suitable Energy Range for Measurement 4

The Low Energy End of the Measurement Range 5

Using a Percentage Signal Strength Metric 5

The Impossibility of Measuring 0% Signal Strength 5

Signal Strength and the Inverse Square Law 5

Experimental Confirmation of Theoretical Assumptions 6

Practical Conversion from Percentage to dBm 7

    Conversion for Atheros 7

    Conversion for Symbol 7

    Conversion for Cisco 8

Conclusions 9

Copyright © 2002 WildPackets, Inc. All Rights Reserved.

Converting Signal Strength Percentage to dBm Values

AiroPeek and AiroPeek NX provide a measurement of RF signal strength represented by a percentage value. The question sometimes arises as to why a percentage metric is used, and how this relates to the actual RF energy that is present in the environment. This paper discusses RF technology with sufficient detail to provide a basis for understanding the issues related to signal strength measurement.

Measurement Units for RF Signal Strength

There are four units of measurement that are all used to represent RF signal strength. These are: mW (milliwatts), dBm ("db"-milliwatts), RSSI (Receive Signal Strength Indicator), and a percentage measurement. All of these measurements are related to each other, some more closely than others. It is possible to convert from one unit to another, albeit with varying degrees of accuracy, and not always in the extremes of the measurement range.

The mW and dBm Units of Measure

The first two units to consider are the mW and the dBm (pronounced "dee-bee-em" or spoken as "dee-bee milliwatts"). When energy is measured in milliwatts (mW), the mW signal level is, simply, the amount of energy present. An electrical engineer or physicist could explain "energy" in more detail, but it is sufficient to think about the fact that a typical wireless access point or quality wireless client NIC has a rated output of 100 mW.

Because of the peculiarities of measurement, it turns out that the measuring RF energy in mW units is not always convenient. This is due, in part, to the fact that signal strength does not fade in a linear manner, but inversely as the square of the distance. This means that if you are a particular distance from an access point and you measure the signal level, and then move twice as far away, the signal will decrease by a factor of four. You move by 2x and the signal decreases by 1/4x; hence, the "inverse square law." In any case, the fact that exponential measurements are involved in signal strength measurement is one reason why the use of a logarithmic scale of measurement was developed as an equivalent, but alternative way of representing RF power.

The "dBm" (dB-milliwatt) is a logarithmic measurement of signal strength, and dBm values can be exactly and directly converted to and from mW values. Just like miles and kilometers can be converted directly, so can mW and dBm (of course, the mW-to-dBm conversion is from a linear scale to a logarithmic scale, and miles-to-kilometers would be linear-to-linear).

A mW measurement is first converted to a base-10 logarithm. It turns out that the logarithm values are quite small; convention multiplies this value by 10 with the resulting value called dBm. Here are some examples to help clarify this relationship:

100mw	log 100 = 2 and 10^2=100	20dBm = 100mW
50mw	log 50 = 1.698 and $10^{1.698}$ = 50	15.9dBm = 50 mW
25mw	log 25 = 1.397 and $10^{1.397}$ = 25	13.9dBm = 25mW
13mw	log 13 = 1.113 and $10^{1.113}$ = 13	11.1dBm = 13mW

You can prove these relationships with your scientific calculator. Notice that each time the actual mW power level becomes half as great, the dBm measurement goes down by (roughly) 3 dBm. As a general guideline, it is convenient to remember that a decrease of 3dBm yields roughly half the original value and, conversely, an increase of 3dBm yields roughly twice the original value.

Of course, power is always a positive quantity. You can't have "negative energy" (unless you're studying quantum mechanics and virtual particles!), so the mW measurement will always be something greater than zero. You can, however, have very small values; much less than 1. When representing a fraction less than 1 (but greater than zero), it can be shown that the corresponding logarithmic value is negative. You can prove the following relationships on your calculator if you desire:

1 mW	log 1 = 0 and 10^0=1	0dBm = 1mW
.5 mW	log .5 = -0.3010 and 10^-0.3010 = .5	-3.01dBm = .5mW
.25mW	log .25 = -0.602 and 10^-0.602=.25	-6.02dBm = .25mW
.13 mW	log .13 = -0.886 and 10^-0.886=.13	-8.86dBm = .13mW

Notice, again, that a decrease of roughly 3dBm yields a change of roughly half in the mW value. It is worth noting, for the discussion that will be presented later in this document, that if you continued building the table until you got down to .0000000002511 mW, you would find that this was equal to –96dBm. It turns out that .0000000002511 mW is about as tiny an RF signal that can be received by most standard 802.11 NICs. This is the "receiver sensitivity" level. As you can see, it is much easier to say, and write, "-96dBm" than to have to figure out where all the zeros are and whether you're talking about "pico-watts" or "fempto-watts" (it is "fempto," by the way; that's why we talk about dBm, which is a much more useful way of measuring signal strength at very low levels).

We can now say, "An 802.11 NIC transmits power at roughly 20dBm and can receive power all the way to –96dBm." You should realize that while it is reasonable to talk about 20dBm as being 100mW, it is cumbersome to talk about –96dBm as being .0000000002511 mW. You should realize that convenience and ease-of-understanding are two fundamental reasons why the dBm metric is used for RF signal strength, rather than mW.

The graph below shows the mathematical relationship between dBm measurements and their corresponding mW values. The actual formula used for the conversion is:

$$dBm = \log(mW) * 10$$

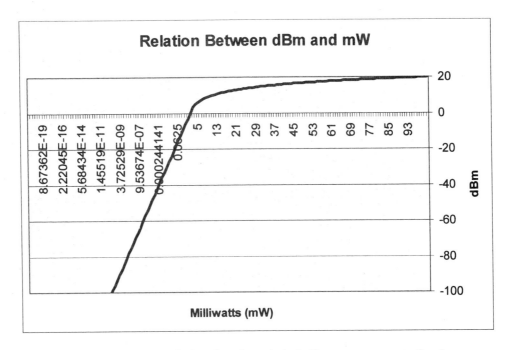

What you can see in the graph is that there is a relatively linear appearance to the slope as it rises from −100 dBm to the point where the mW value is roughly 5. At that point, the curve turns sharply to the right and flattens out. After the curve a relatively large change in mW value is required to make a significant change in dBm. This is purely a mathematical relationship, however. As will be shown, the fact that dBm measurements don't change much above 5 mW will be significant in how 802.11 NIC manufacturers have chosen to present RF signal strength measurements.

The Receive Signal Strength Indicator (RSSI)

The IEEE 802.11 standard defines a mechanism by which RF energy is to be measured by the circuitry on a wireless NIC. This numeric value is an integer with an allowable range of 0-255 (a 1-byte value) called the Receive Signal Strength Indicator (RSSI). No vendors have chosen to actually measure 256 different signal levels, and so each vendor's 802.11 NIC will have a specific maximum RSSI value ("RSSI_Max"). For example, Cisco chooses to measure 101 separate values for RF energy, and their RSSI_Max is 100. Symbol uses an RSSI_Max value of 31. The Atheros chipset uses an RSSI_Max value of 60. Therefore, it can be seen that the RF energy level reported by a particular vendor's NIC will range between 0 and RSSI_Max. Notice that nothing has been said here about measurement of RF energy in dBm or mW. RSSI is an arbitrary integer value, defined in the 802.11 standard and intended for use, internally, by the microcode on the adapter and by the device driver. For example, when an adapter wants to transmit a packet, it must be able to detect whether or not the channel is clear (i.e., nobody else is transmitting). If the RSSI value is below some very low value, then the chipset knows that the channel is clear. This is the "Clear Channel Threshold" and some particular RSSI value is associated with it. When an 802.11 client is associated to an access point and is roaming, there comes a point when the signal level received from the access point drops to a somewhat low value (because the client is moving away from the access point). This level is called the "Roaming Threshold" and some intermediate (but low) RSSI value is associated with it. Different vendors use different signal levels for the Clear Channel Threshold and the Roaming Threshold and, moreover, the RSSI value that represents these thresholds differs from vendor-to-vendor because different RSSI_Max values are implemented.

RSSI in the 802.11 Standard

Here is what the IEEE 802.11 standard says about the RSSI metric:

14.2.3.2 RXVECTOR RSSI

The receive signal strength indicator (RSSI) is an optional parameter that has a value of 0 through RSSI Max. This parameter is a measure by the PHY sublayer of the energy observed at the antenna used to receive the current PPDU. RSSI shall be measured between the beginning of the start frame delimiter (SFD) and the end of the PLCP header error check (HEC). RSSI is intended to be used in a relative manner. Absolute accuracy of the RSSI reading is not specified.

Notice that the parameter is specified as optional, although all 802.11 NIC manufacturers appear to implement it. Of greatest significance are the last two sentences: "The RSSI is intended to be used in a relative manner. Absolute accuracy of the RSSI reading is not specified."

There is no specified accuracy to the RSSI reading. That is, there is nothing in the 802.11 standard that stipulates a relationship between RSSI value and any particular energy level as would be measured in mW or dBm. Individual vendors have chosen to provide their own levels of accuracy, granularity, and range for the actual power (measured as mW or dBm) and their range of RSSI values (from 0 to RSSI_Max).

Granularity in RSSI Measurements

The concept of "granularity" is important to consider here, too. Since the RSSI value is an integer it must increase or decrease in integer steps. For example, Symbol provides 32 separate "steps," Cisco provides 101 (i.e., from 0 to RSSI_Max for any given manufacturer). Whatever range of actual energy is being measured, it must be divided into the number of integer steps provided by the RSSI range. Therefore, if RSSI changes by 1, it means that the power level changed by some proportion in the measured power range. There are, therefore, two important considerations in understanding RSSI. First, it is necessary to consider what range of energy (the mW or dBm range) that's actually being measured. Secondly, it must be recognized that all possible energy levels (mW or dBm values) cannot be represented by the integer set of RSSI values.

The Choice of a Suitable Energy Range for Measurement

As was seen in the dBm-to-mW graph above, there is not much change in dBm values above roughly 5 mW. Wireless NIC manufacturers do not measure signal strength in that range. RF energy is almost always measured using dBm values, because the measured range would otherwise have mW values with too many zeros to the right of the decimal point to make for ease of understanding. The graph also shows that the slope of change for dBm below 5mW is very roughly linear, but not exactly. The logarithmic nature of the dBm measurement, coupled with the fact that the RSSI range used for measurement contains dBm "gaps" (due to the integer nature of the RSSI value), has led many vendors to map RSSI to dBm using a table. These mapping tables allow for adjustments to accommodate the logarithmic nature of the curve. The range of energy that is typically measured begins at or below –10dBm (and, compared to the +20dBm of potential output power at a 100mW access point, that's a relatively weak signal). In addition to the fact that the graph "flattens out" at higher power levels, the –10dBm upper limit on the energy level measurement range is also consistent with the purpose for RSSI measurements in the first place. Remember that RSSI is intended for use in Clear Channel assessment and determination of the Roaming Threshold. It makes sense that the circuitry is designed to provide reasonable accuracy in this range.

The Low Energy End of the Measurement Range

The receiver circuit in an 802.11 NIC must have a minimum level of available RF energy (above the level of the background noise) in order to extract a bit-stream. This minimum level is called the "Receive Sensitivity" and is a NIC spec measured in dBm. For example, a NIC manufacturer may indicate that their particular card has a Receive Sensitivity of –96dBm at 1Mb/sec. If the actual RF energy present at that card were less than –96dBm then the card would no longer be able to differentiate between signal and noise. The dBm value for a NICs Receive Sensitivity is very close to the dBm value associated with an RSSI value of 0. Hence, the Receive Sensitivity of the adapter determines the lower end of the necessary measurement range for signal strength. It should be noted that, typically, if RSSI=0 the dBm signal measurement is below the Receive Sensitivity level.

Using a Percentage Signal Strength Metric

To circumvent the complexities (and potential inaccuracies) of using RSSI as a basis for reporting dBm signal strength, it is common to see signal strength represented as a percentage. The percentage represents the RSSI for a particular packet divided by the RSSI_Max value (multiplied by 100 to derive a percentage). Hence, a 50% signal strength with a Symbol card would convert to an RSSI of 16 (because their RSSI_Max = 31). Atheros, with RSSI_Max=60, would have RSSI=30 at 50% signal strength. Cisco ends up making life easy with an RSSI_Max =100, so 50% is RSSI=50.

It can be seen that use of a percentage for signal strength provides a reasonable metric for use in network analysis and site survey work. If signal strength is 100%, that's great! When signal strength falls to roughly 20%, you're going to reach the Roaming Threshold. Ultimately, when signal strength is down somewhere below 10% (and probably closer to 1%), the channel is going to be assumed to be clear. This conceptualization obviates the need to consider dBm, the RSSI_Max, or the "knee" in the logarithmic curve of mW to dBm conversion. It allows a reasonable comparison between environments even though different vendor's NICs were used to make the measurements. Ultimately, the generalized nature of a percentage measurement allows the integer nature of the RSSI to be overlooked.

The Impossibility of Measuring 0% Signal Strength

In the preceding paragraph, you may have noticed the note in parentheses stating that the clear channel threshold was "probably closer to 1%." There is a very profound reason why this was not "0%." If signal strength falls to 0%, it can be assumed that RSSI=0 and, hence, the signal strength is at, or below, the Receive Sensitivity of the NIC. A NIC can't report that a particular packet has "0% signal strength," because if there were no available signal, there would be no packet to measure!

It is impossible for any tool using a standard wireless NIC to measure signal strength below the NIC's Receive Sensitivity threshold.

Signal Strength and the Inverse Square Law

Earlier, it was stated that the "inverse square law" defined how an RF signal would be reduced in power. A physicist might explain that there are other factors that come into play with signal attenuation, but the inverse square law has the most dramatic impact. In fact, when measurements are taken at a distance greater than approximately one wavelength away from an electromagnetic radiator, the other influences to the energy level of the radiated wave become insignificant and can be ignored. Imagine that a 100mW access point actually had 100mW of measured power 1-inch away from the antenna. Of course, this is a thought experiment only. Antenna loss or gain, and the actual energy of the radiated signal, would influence the real-world measured power and probably not be exactly 100mW. The thought experiment is, nonetheless, interesting. Immediately we're faced with an impossible situation. If the measured power were 100mW at a distance of 1-inch, then the measured power would be 400mW at a distance of ¾-inch (by the inverse-square law). At ¼-inch, the power would have to jump up to 1600mW. Obviously, this is not reality. In fact, theoretical measurements of RF signals are a challenge for students in college physics classes and they involve some complicated formulae. You see, when a transmitter is rated at 100mW, there is an implication that a 100mW signal is present at the last point in the transmitter circuit before the signal enters the antenna! The antenna will introduce some loss or gain, and what comes out will be assumed to be at the power level derived from adjusting the power at the antenna by that gain or loss.

Continuing the thought experiment, however, and recognizing that real-world measurement would probably be smaller than those derived in the thought experiment, adds to an understanding of why RSSI values are associated with dBm signal strengths at levels below -10dBm. If the measured power at 1-inch from an antenna were 100mW, then we could imagine the following measurements, based on the inverse-square law:

1" = 100mW = 20dBm
2" = 25mW = 13.9dBm
4" = 6.25mW = 7.9dBm
8" = 1.56mW = 1.9dBm
16" = 0.39mW = -4.08dBm
32" = .097mW = -10.1dBm
64" = .024mW = -16.1dBm (5.3 feet away)
128" = .006mW = -22.2dBm (10.6 feet away)
256" = .0015mW = -28.2dBm (21.3 feet away)

What you see from the table is that somewhere between roughly 5 feet and 20 feet away from a 100mW radiator, the signal strength falls to below -20dBm. Consequently, measurements represented by RSSI values that refer to energy levels below -10dBm (or lower) are reasonable and practical.

Experimental Confirmation of Theoretical Assumptions

Engineers at WildPackets performed some limited experiments to see how closely these theoretical concepts matched with real-world measurements. An access point rated at 100mW was measured using AiroPeek NX. It was found that within 5 feet of the access point, the indicated signal strength remained at 100%, indicating that the signal was as strong or stronger than the high-end of the dBm range of measurement used by the NIC. Between 5 and 10 feet away from the access point the signal strength occasionally fell to as low as 80%, but essentially hovered at 100%, with only occasional drops to 80%. Measurements were made on the far side of a drywall-on-wooden-stud wall. It was found that there, too, signal strength remained at 100% within 5-feet of the access point. The conclusion drawn from these experiments was that use of AiroPeek for measuring signal strength was reasonable beyond 10 feet from the access point. However, the variations in measured signal level (the "hovering" of the signal strength with a 20% variation) meant that site survey measurements would be generalized, at best. General measurements would be suitable for many practical site survey situations, since a key determination in a site survey involves identifying places where the signal strength is unacceptably low, as opposed to creating an accurate dBm signal strength map of a particular environment. As long as the measured signal strength remains above 30%, there should be sufficient signal for normal 802.11 operations. In practice, one could determine the signal strength percentage at which 802.11 speed dropped from 11Mb/sec to 5.5Mb/sec, and then the level at which the rate dropped to 2Mb/sec or 1Mb/sec, and use those determined signal strength percentages as part of a site survey baseline.

Practical Conversion from Percentage to dBm

The effectiveness or reasonability of using dBm measurements obtained from a standard wireless NIC is questionable when used as part of a real-world network troubleshooting exercise. This is because most NICs only provide RSSI in a range that is below −10dBm, and everything above that is mapped to RSSI_Max (or, 100% signal strength). If it is important to know the difference between −40dBm and −50dBm, then why isn't it equally important to know the difference between +20dBm and +10dBm? Moreover, is it not important to determine the actual output power of an antenna, particularly a directional antenna? The output power would be measured in positive dBm, possibly even greater than +20dBm (for a high-gain antenna). These measurements are outside the RSSI range for most adapters.

Nonetheless, following are conversion tables, based on information obtained from various NIC manufacturers, which will provide a mapping between RSSI and dBm. There is a two-step process when going from a percentage signal strength report in an analyzer to the dBm value in a vendor's table. First, it is necessary to know the RSSI_Max for the vendor and, from that, the RSSI that corresponds to the current percentage value can be obtained (i.e., x% of RSSI_Max = RSSI). Once the RSSI value has been obtained from the percentage, it is only necessary to plug it in to the vendor's table (or formula) and get a dBm value. You should notice, in each description that follows, how the values in the tables don't always increase in a linear manner. Sometimes a table value will go up by 5, other times by 6, and so forth. This is to account for the logarithmic nature of dBm measurements. Embodied in these "gaps" in the table, and exacerbated by the integer nature of the RSSI, are inherent potential inaccuracies that must be recognized.

Conversion for Atheros

Unlike the other vendors described, Atheros uses a formula to derive dBm.

RSSI_Max = 60

Convert % to RSSI

Subtract 95 from RSSI to derive dBm

Notice that this gives a dBm range of −35dBm at 100% and −95dBm at 0%.

Conversion for Symbol

RSSI_Max = 31

Convert % to RSSI and lookup the result in the following table:

RSSI <= 4 is considered to be −100dBm

RSSI <=8 is considered to be −90 dBm

RSSI <=14 is considered to be −80 dBm

RSSI <=20 is considered to be −70 dBm

RSSI <=26 is considered to be −60 dBm

RSSI greater than 26 is considered to be −50dBm

Notice that this gives a dBm range of −50dBm to −100dBm but only in 10dBm steps.

Conversion for Cisco

Cisco has the most granular dBm lookup table.

RSSI_Max = 100

Convert % to RSSI and lookup the result in the following table. The RSSI is on the left, and the corresponding dBm value (a negative number) is on the right.

RSSI	dBm	RSSI	dBm	RSSI	dBm
0	= −113	34	= −78	68	= −41
1	= −112	35	= −77	69	= −40
2	= −111	36	= −75	70	= −39
3	= −110	37	= −74	71	= −38
4	= −109	38	= −73	72	= −37
5	= −108	39	= −72	73	= −35
6	= −107	40	= −70	74	= −34
7	= −106	41	= −69	75	= −33
8	= −105	42	= −68	76	= −32
9	= −104	43	= −67	77	= −30
10	= −103	44	= −65	78	= −29
11	= −102	45	= −64	79	= −28
12	= −101	46	= −63	80	= −27
13	= −99	47	= −62	81	= −25
14	= −98	48	= −60	82	= −24
15	= −97	49	= −59	83	= −23
16	= −96	50	= −58	84	= −22
17	= −95	51	= −56	85	= −20
18	= −94	52	= −55	86	= −19
19	= −93	53	= −53	87	= −18
20	= −92	54	= −52	88	= −17
21	= −91	55	= −50	89	= −16
22	= −90	56	= −50	90	= −15
23	= −89	57	= −49	91	= −14
24	= −88	58	= −48	92	= −13
25	= −87	59	= −48	93	= −12
26	= −86	60	= −47	94	= −10
27	= −85	61	= −46	95	= −10
28	= −84	62	= −45	96	= −10
29	= −83	63	= −44	97	= −10
30	= −82	64	= −44	98	= −10
31	= −81	65	= −43	99	= −10
32	= −80	66	= −42	100	= −10
33	= −79	67	= −42		

Notice that this gives a range of −10dBm to −113dBm. Bearing in mind that a Cisco card will have a Receive Sensitivity of −96dBm at its lowest, it is impossible to obtain an RSSI value of less than 16. Note, also, that all RSSI values greater than 93 are assigned −10dBm, and that there are multiple places in the table where two adjacent RSSI values are assigned the same dBm value.

There is another aspect to interpreting the Cisco RSSI. The Cisco device driver code indicates that if the RSSI value converts to less than –90dBm, then it should be converted to a fixed value of –75dBm. This leaves a question as to the exact interpretation of an RSSI value that converts to –76dBm to –89dBm. A measurement of –75dBm would be reported as 36% signal strength. It seems sufficient to leave this quandary unanswered, since whether a signal is at –75dBm or at –92dBm, the entire low-end range is less than what would be desirable for normal 802.11 WLAN operation.

Conclusions

While it is possible to arrive at a dBm value that is somewhat equivalent to a reported Signal Strength percentage, the absolute accuracy of the value is questionable. The range of possible measurements is below –10dBm, which precludes using an off-the-shelf NIC for measurement of high-gain antennae, or anywhere close to an access point (where the signal level is above –10dBm). In general, the use of a percentage value for signal strength allows for a relatively simple, consistent, reproducible metric that can be used as part of a site survey. When, however, accurate dBm measurements must be made as part a site survey, or in the course of network analysis and troubleshooting, the use of an RF Spectrum Analyzer tool should be considered.

In practice, it can be observed that above some particular signal strength (%) traffic moves at 11Mb/sec (in 802.11b) and, as the percentage decreases, there's a point where the data begins moving at 5.5 Mb/sec. Still later the speed drops, ultimately, to 1 Mb/sec, and finally there are increased numbers of CRC errors with an ultimate loss of reception. Real-world testing can establish the signal levels (as percentages) associated with each of these events, and these can then be used as a baseline for ongoing measurement and analysis.

WildPackets Professional Services

WildPackets offers a full spectrum of unique professional support services, available on-site, online or through remote dial-in service.

WildPackets Academy

WildPackets Academy provides the most effective and comprehensive network and protocol analysis training available, meeting the professional development and training requirements of corporate, educational, government, and private network managers. Our instructional methodology and course design centers around practical applications of protocol analysis techniques for Ethernet and 802.11 wireless LANs.

In addition to classroom-taught Network Analysis Courses, WildPackets Academy also offers:
- Web-Delivered Training
- On-site and Custom Courseware Delivery
- The Technology, Engineering, and Networking Video Workshop Series
- On-site and Remote Consulting Services
- Instruction and testing for the Network Analysis Expert (NAX™) Certification

For more information about consulting and educational services, including complete course catalog, pricing and scheduling, please visit www.wildpackets.com/academy. NAX examination and certification details are available at www.nax2000.com.

Live Online Quick Start Program

WildPackets now offers one-hour online Quick Start Programs on using EtherPeek NX/EtherPeek and AiroPeek NX/AiroPeek, led by a WildPackets Academy Instructor. Please visit www.wildpackets.com for complete details and scheduling information.

About WildPackets, Inc.

WildPackets, a privately-held corporation, was founded in 1990 with a mission to create software-based tools to simplify the complex tasks associated with maintaining, troubleshooting, and optimizing evolving computer networks. WildPackets' patented, core "Peek" technology is the development base for EtherPeek™, TokenPeek™, AiroPeek™, and the NX™ family of expert packet analyzers. All are recognized as the analysis tools of choice for small, medium, and large enterprise customers, allowing IT Professionals to easily maximize network productivity. Information on WildPackets, WildPackets Academy, Professional Services, products, and partners is available at www.wildpackets.com.

WildPackets, Inc.
1340 Treat Blvd, Suite 500
Walnut Creek, CA 94597
925-937-3200
www.wildpackets.com

20021217-M-WP007